RIVERS, LAKES, STREAMS, AND PONDS

Richard Beatty

Steck-Vaughn Company

First published 2003 by Raintree Steck-Vaughn Publishers,
an imprint of Steck-Vaughn Company.
Copyright © 2003 The Brown Reference Group plc

Library of Congress Cataloging-in-Publication Data

Beatty, Richard.
  Rivers, lakes, streams, and ponds / Richard Beatty.
    v. cm. -- (Biomes atlases)
  Contents: Biomes of the world -- Rivers and lakes of the world -- Still
and flowing waters -- Water plants -- Animal life -- Rivers, lakes, and
people -- The future.
  ISBN 0-7398-5513-1 (lib. bdg. : hardcover)
  1. Freshwater ecology--Juvenile literature. 2. Freshwater
ecology--Maps--Juvenile literature. [1. Freshwater ecology. 2.
Ecology.] I. Title. II. Series.

QH541.5.F7 B43 2002
577.6--dc21   2002012817

Printed in Singapore. Bound in the United States.
1 2 3 4 5 6 7 8 9 0 LB 07 06 05 04 03 02

The Brown Reference Group plc
Project Editor: Ben Morgan
Deputy Editor: Dr. Rob Houston
Consultant: William Kleindl, Aquatic Ecologist,
       Paramatrix Inc., Seattle, WA
Designer: Reg Cox
Cartographers: Mark Walker and
       Darren Awuah
Picture Researcher: Clare Newman
Indexer: Kay Ollerenshaw
Managing Editor: Bridget Giles
Design Manager: Lynne Ross
Production: Alastair Gourlay

Raintree Steck-Vaughn
Editor: Walter Kossmann
Production Manager: Brian Suderski

Front cover: Tributary of the Amazon River.
Inset: Mallard ducklings.

Title page: A stream in the Tasmanian
temperate rain forest, Australia.

The acknowledgments on p. 64 form
part of this copyright page.

About this Book

This book's introductory pages describe the biomes of the world and then the river and lake biomes. The five chapters look at aspects of rivers, lakes, streams, and ponds: climate, plants, animals, people, and future. Between the chapters are detailed maps that focus on important rivers and lakes. The map pages are shown in the contents in italics, **like this**.

Throughout the book you'll also find boxed stories or fact files about rivers and lakes. The icons here show what the boxes are about. At the end of the book is a glossary, which explains what all the difficult words mean. After the glossary is a list of books and websites for further research and an index, allowing you to locate subjects anywhere in the book.

 Climate

 People

 Plants

 Future

 Animals

 Facts

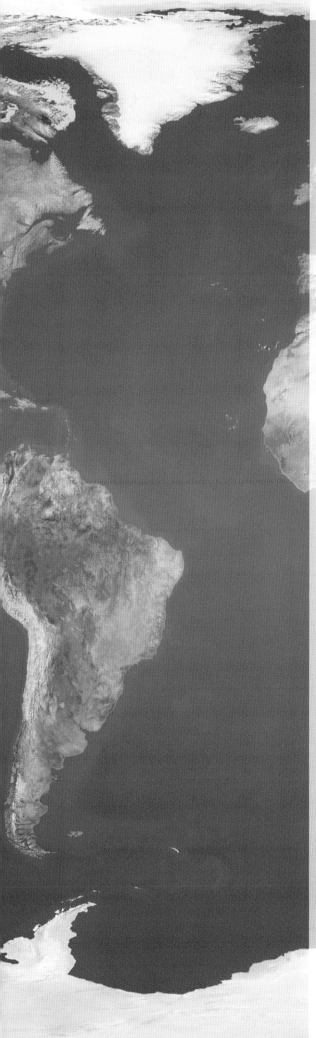

Contents

Biomes of the World

Biologists divide the living world into major zones named biomes. Each biome has its own distinctive climate, plants, and animals.

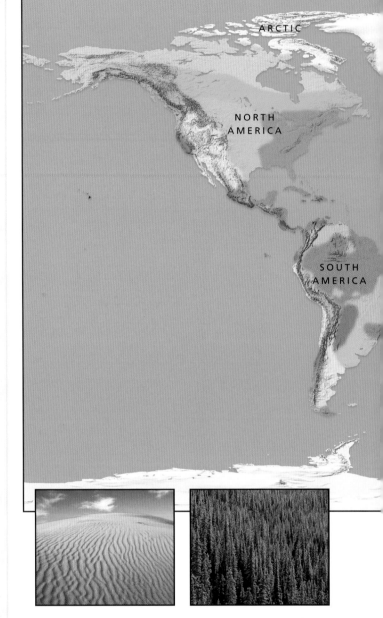

Desert is the driest biome. There are hot deserts and cold ones.

Taiga is made up of conifer trees that can survive freezing winters.

If you were to walk all the way from the north of Canada to the Amazon rain forest, you'd notice the wilderness changing dramatically along the way.

Northern Canada is a freezing and barren place without trees, where only tiny brownish-green plants can survive in the icy ground. But trudge south for long enough and you enter a magical world of conifer forests, where moose, caribou, and wolves live. After several weeks, the conifers disappear, and you reach the grass-covered prairies of the central United States. The farther south you go, the drier the land gets and the hotter the sun feels, until you find yourself hiking through a cactus-filled desert. But once you reach southern Mexico, the cacti start to disappear, and strange tropical trees begin to take their place. Here, the muggy air is filled with the calls of exotic birds and the drone of tropical insects. Finally, in Colombia you cross the Andes mountain range—whose chilly peaks remind you a little of your starting point—and descend into the dense, swampy jungles of the Amazon rain forest.

Scientists have a special name for the different regions—such as desert, tropical rain forest, and prairie—that you'd pass through on such a journey. They call them biomes. Everywhere on Earth can be classified as being in one biome or another, and the same biome often appears in lots of

4

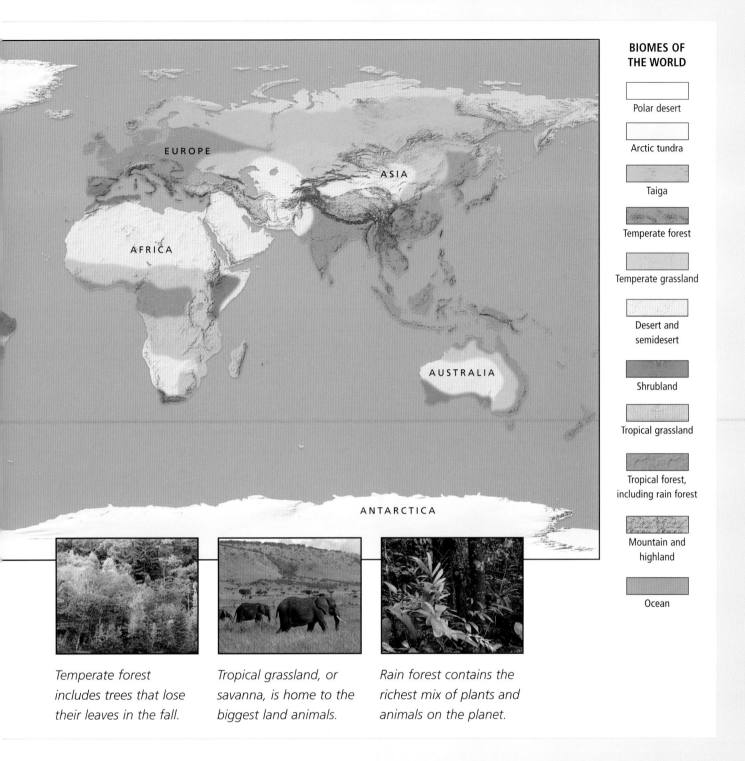

BIOMES OF THE WORLD

Polar desert

Arctic tundra

Taiga

Temperate forest

Temperate grassland

Desert and semidesert

Shrubland

Tropical grassland

Tropical forest, including rain forest

Mountain and highland

Ocean

EUROPE

ASIA

AFRICA

AUSTRALIA

ANTARCTICA

Temperate forest includes trees that lose their leaves in the fall.

Tropical grassland, or savanna, is home to the biggest land animals.

Rain forest contains the richest mix of plants and animals on the planet.

different places. For instance, there are areas of rain forest as far apart as Brazil, Africa, and Southeast Asia. Although the plants and animals that inhabit these forests are different, they live in similar ways. Likewise, the prairies of North America are part of the grassland biome, which also occurs in China, Australia, and Argentina. Wherever there are grasslands, there are grazing animals that feed on the grass, as well as large carnivores that hunt and kill the grazers.

The map on this page shows how the world's major biomes fit together to make up the biosphere—the zone of life on Earth.

Rivers and Lakes of the World

Rivers, lakes, and other inland waters form one of the world's most varied biomes. From icy pools in the Arctic to enormous inland seas, and from mountain torrents to tropical rivers miles across, they provide a range of habitats for animals and plants.

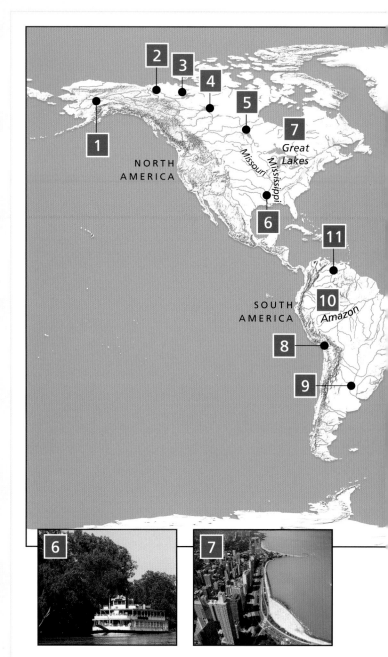

Barges and pleasure boats ply the great Mississippi waterway.

Cities such as Chicago have grown up on the Great Lakes' shores.

The animals and plants of rivers and lakes do not live in the most stable of environments. Rivers and ponds may dry up altogether, while floods may alter rivers and valleys beyond recognition. Nonetheless, a great variety of creatures survive in this biome. Some 40 percent of all species of fish live in freshwater, although rivers and lakes cover a tiny area compared to oceans.

Many great rivers start in mountains and highlands, such as the Chang (Yangtze) in China, which flows from the Tibetan Plateau, and the Ganges River of India, flowing from the Himalayas. Running down from the mountains, a river might reach any kind of landscape, from arctic tundra to baking desert. Those that flow through rain forest regions are fed by plentiful tropical rain and, like the Amazon and Congo rivers, become vast waterways miles across.

Wherever a depression in the land causes water to collect, a lake forms. Like rivers, most lakes are freshwater, and their animals and plants differ from those of the great

6

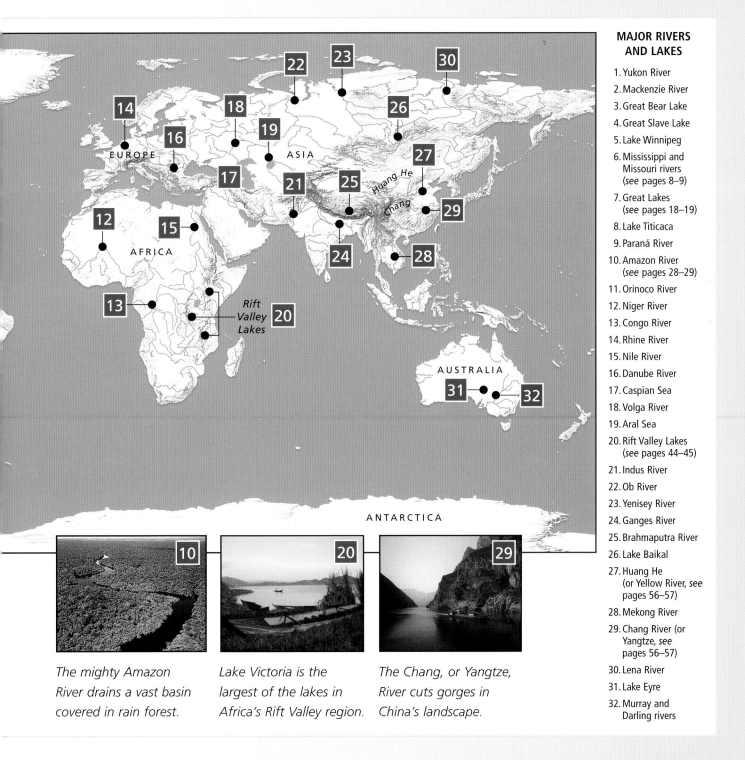

The mighty Amazon River drains a vast basin covered in rain forest.

Lake Victoria is the largest of the lakes in Africa's Rift Valley region.

The Chang, or Yangtze, River cuts gorges in China's landscape.

saltwater biome, the ocean. Many lakes are salty, though—sometimes much saltier than the sea, supporting only a small selection of unusual creatures.

Rivers and lakes are unique in the way they affect nearby biomes. Rivers in flood vary the types of plants in valleys by creating marshes and swamps, while a desert stream may support the only trees for miles. Rivers can act as corridors for the movement of fish and waterbirds, but they can be barriers to land animals. However, it is aquatic wildlife that is truly isolated: It is a real challenge to travel to new rivers and lakes across barriers of land.

7

Mississippi

Forming the largest river system in North America, the Mississippi and its tributaries have played a central role in the history of the United States.

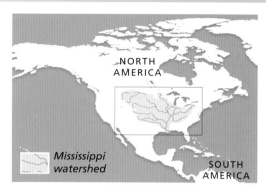

Mississippi watershed

NORTH AMERICA

SOUTH AMERICA

 ## Fact File

▲ After the last ice age, the Mississippi Valley allowed fish and other freshwater creatures to recolonize much of North America from the south, which had remained ice free.

▲ The Mississippi system has the greatest diversity of freshwater mussels of any river system, with more than sixty species. At one time exploited for the pearl button industry, these creatures are currently under threat from pollution, clogging by sediment, and alterations to river flow by dams.

Engineers have built many structures across the river. Some generate electricity, others control flooding, but all of them obstruct the movement of wildlife.

 ## Dammed Floods

The Mississippi River and its tributaries have a long history of human intervention. Before Europeans settled there, Native Americans cleared some of the surrounding land for farming. European agriculture later led to many tributaries being clogged with soil washed from newly plowed land. In the late 19th century, dams were built on the upper Mississippi to help ships pass more easily. Later, huge dams, mainly for hydropower, were also built on the Tennessee River and the Missouri. Today, generators in dams on the Tennessee River produce enough hyrdoelectricty to power 1.5 million homes. There are so many dams that wildlife can no longer migrate freely up and down the river.

The Mississippi's flow is also modified with large structures that control flooding. Since the disastrous floods of 1993, however, planners are considering natural flood control—by restoring wetlands.

An Alabama hog sucker (right) grazes algae from rocks in the clear, fast-running waters of the Tennessee river system. Young mussels live on the fish as parasites.

1. Missouri River
The Missouri and its major tributary, the Yellowstone, both carry a lot of sediment from the Rocky Mountains and are responsible for the muddiness of the lower Mississippi.

2. Lake Itasca
This lake and the streams flowing into it are the source of the Mississippi River.

3. Chicago Sanitary and Ship Canal
Joining Lake Michigan and the Illinois River, this canal allows ships to pass between the Mississippi and the Great Lakes but has also allowed invasive pests such as the zebra mussel to spread more easily.

4. Cahokia
Once one of the largest centers of the Mississippian civilization, this ancient city flourished more than 800 years ago.

5. St. Louis
Despite being protected by high riverside banks (levees), St. Louis almost suffered disastrous flooding in the 1993 Mississippi floods. It was saved by a levee giving way on the far riverbank.

6. Bottomlands
The low-lying land where the Missouri and Mississippi meet was covered by distinctive forests and swamps. Most of these have been cleared, mainly for growing cotton, soybeans, rice, and corn in the fertile soil.

7. Ohio River
This large tributary has suffered pollution from industrial activity in its basin near cities such as Pittsburgh and Cincinnati.

8. Appalachian, Ozark, and Ouachita Highlands
Stretches of river unmodified by dams in these highlands are rich in wildlife. Here live many species of snails, mussels, and crayfish and, in the larger rivers, ancient types of fish such as sturgeons and paddlefish.

9. Mississippi Delta
The Mississippi has changed its outlet to the sea several times in the last few thousand years. Each former outlet is shown by a lobe of the present delta.

10. Dead Zone
Pollution in the Mississippi causes the oxygen content of its waters to be low. Where the oxygen-poor Mississippi water enters the ocean, it causes a dead zone, where marine life cannot flourish.

Still and Flowing Waters

From whitewater rapids to placid lakes, rivers and lakes create a wealth of habitats for wildlife. They also play a vital role in Earth's water cycle and help shape the landscape.

Without the heat from the sun driving the world's weather systems, Earth's land surface would be dry and lifeless. The water that evaporates (turns to vapor) from the ocean later falls as rain or snow and makes life on land possible. Some of the water soaks deep into the ground. Another fraction of the water is taken up by plants or evaporates again. But much of it finds its way into rivers, the land's natural drains. It flows downhill until it reaches the lowest level possible—usually the sea, but sometimes an inland lake. Worldwide, this sequence of events is known as the water cycle.

Rivers vary greatly. Some are fed by a very small area—the side of a snowcapped mountain, for example—while others collect water from a vast area of land. The area that a river drains is called its watershed. In a large watershed, there may be dozens of small rivers and streams, or tributaries, flowing into the main river. Together, these rivers and streams make up a river system.

The Amazon river system has the biggest watershed of all, covering about one-third of South America. The watershed of the Mississippi river system is nearly as big, but this system carries only about one-tenth as much water as the Amazon because far less rain falls there than in South America.

Some rivers, such as the Rhine River in Europe, flow fairly evenly throughout the year. Others have annual floods, driven either by the spring thaw in cooler climates or by a rainy season. Some rivers even freeze over during winter. Rivers also vary a lot in the amount of sediment (sand and mud particles) they carry. The Mississippi and the Amazon are both very muddy rivers, for example. Many of their fish find their way using smell and touch rather than sight.

The Long Term

Rivers slowly wear away the ground and, over millions of years, reshape the landscape. High mountains and highlands, pushed up by

 Fact File

▲ Lakes are not lakes forever: Even large lakes fill slowly with sediment and disappear. By far the oldest is Lake Baikal in Siberia, at 25 million years old. Most lakes are less than 12,000 years old.

▲ Waterfalls are natural barriers to fish and other animals that cannot reach the river or lake habitat above a cascade. Wildlife must now face new obstructions—dams and other artificial structures.

▲ Lakes are isolated from one another. Lake plant and animal species must travel from lake to lake or they will become extinct when the lake disappears.

The ancient, hard rock of the Grand Teton range in Wyoming
(background) is not easily worn away by streams and rivers.
As a result, the water of the Snake River (foreground) is
often clear and free of fine sediment.

forces deep inside Earth, are ground down by glaciers, split by frost, or worn down by rivers and wind. Streams and rivers carry away the crumbling material, carving out valleys as they do so. New mountains form as old ones wear away, so the cycle continues.

The result of these ongoing processes is a patchwork of watersheds, each drained by a different river system and separated by higher ground. In the United States, the Rocky Mountains form a barrier that separates rivers draining westward toward the Pacific from those flowing east into the Mississippi. For aquatic (water-dwelling) animals, crossing from one watershed to another may be impossible. As a result, many river systems contain unique species that evolved within them and that live nowhere else.

A river system may have a complex history. Millions of years ago, for example, sea levels rose and dammed back the water of the Amazon river system, so much of the watershed turned into a kind of swampy lake. Later, during the Ice Age (a period when Earth's climate became much colder), sea levels were much lower than today, and the

River water ranges from clear to muddy and from energetic to sluggish. Chinook salmon (above) must swim against the raging torrents of mountain rivers, dodging waiting brown bears. Catfish (below) have their own problems—to find their way through murky water they need long, sensitive barbels (fleshy whiskers).

Amazon's river valleys became deep ravines. Today, these canyons are invisible, drowned by sediment or filled with water more than 300 feet (90 m) deep.

Sometimes a river keeps cutting down into its valley while the land around is rising; the Grand Canyon in Arizona is the most spectacular result of this process. In other regions, such as the Great Lakes, the whole area was buried under solid ice during the Ice Age. The animals now living in the region arrived during the last 15,000 years, after the ice finally melted.

Compared to rivers, which flowed through watersheds long before the Ice Age, most lakes are relatively young features of Earth's

formed this way. However, some of the world's oldest and deepest lakes have a different origin. In areas such as East Africa, movements in Earth's crust created vast splits in the land that filled with water and formed deep lakes; some are now several million years old. A lake can also form in a valley that has been blocked by a landslide, in the crater of a volcano, or occasionally in an old meteor crater.

Rushing Waters

Rivers and streams provide many different kinds of habitats for animals and plants. The way the water flows has a huge influence on the kinds of creatures that can survive in it. Most rivers start as swift-flowing streams among hills or mountains. The churning rapids capture plenty of oxygen from the air, helping fish and other aquatic animals breathe underwater. Getting a meal, though, can be a challenge. Few plants can avoid being swept away by the powerful current— and in any case, there is

surface. They tend to fill in with sediment brought by rivers. Most of the world's lakes were created by glaciers during the Ice Age. Glaciers are slow-flowing rivers of ice that scour great hollows out of the land as they move. North America's Great Lakes

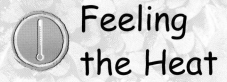

Feeling the Heat

You might not think that anything could survive in the hot, salty springs of volcanic areas such as Yellowstone Park (left), but a few organisms can. Bacteria and algae form the base of food chains, and insects such as water beetles and midge larvae also thrive in the springs. There are even tiny worms that can stand temperatures of 140°F (60°C)—more than halfway to the boiling point of water.

usually no mud in which to take root. And without plants, there can be no plant-eating animals for the fish to eat.

The biggest source of food in a stream is not plants but waste material from outside the stream. Especially where a stream runs through a forest, dead leaves and rotting wood fall into the water and form the starting point for a community of living things, beginning with fungi and bacteria and

A stream that flows through a tropical forest picks up a steady supply of dead plant parts, dead animals, and feces from above. All the animals in the stream rely on this material, directly or indirectly, for food.

other microorganisms that break down the dead material. These organisms, which form the slime around rocks and leaves in streams, are the main food source for aquatic larvae (young stages) of insects. Mayfly larvae, for

World of Darkness

A river can disappear underground wherever a hole or crack lets the water through. It may continue flowing underground, and over thousands of years, it can carve a system of caves through the rock. Such caves might seem cut off from the life-giving light and warmth of the sun, but life continues in the darkness. Cave animals live on morsels of food carried by the river water from the sunlit biomes above. The top predators in caves are often blind cave fish (left).

instance, scrape the slime off rocks and eat it but in turn become a meal for predators such as water beetles, water bugs, and fish.

Flowing Smoothly

As a stream or river reaches lower ground, its slope gets gentler and the water flow becomes smoother. It may still flow fast, but near the riverbed there is calmer water that allows sand and mud to settle. Animals such as worms and mussels bury themselves in the bed and filter mud and water for small food particles. Waterweeds may take root, trailing their long stems in the water, while tall plants such as reeds often grow around the water's edge. Other rivers are too deep or muddy for underwater plants to grow.

Some rivers fall from great heights, forming thundering waterfalls. A large waterfall can create its own mini-environment, where the air is constantly filled with mist and spray. It can also provide hiding places: South America's massive Iguaçú Falls shelter a type of swift that can dive straight through the curtain of water with its wings folded before finding a safe perch on the rock behind.

A river nearing the sea usually slows and deposits sediment. Banks of sediment can serve as turtle nesting sites or as places for crocodiles to beach themselves.

Many rivers, such as the Mississippi, flood in their lower course, forming a patchwork of marshes, shallow lakes, and other habitats.

Like other waterfalls, Iguaçú Falls in southern Brazil form a barrier that many river animals cannot cross.

The Amazon River floods a huge area of forest every year; the trees have evolved to survive being flooded for months at a time. Many fish swim into the forest at this time to feed and reproduce. Some trees drop their fruit during the flood and depend on fish to eat and distribute the seeds.

Eventually, the river reaches the sea. The mixing of salt water and freshwater, the rise and fall of the tides, and the muddy particles that settle on the river bottom create a variety of special habitats. Broad mudflats may glisten at low tide, hiding countless

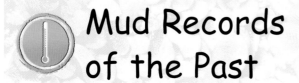

Mud Records of the Past

Scientists can find out a lot by studying sediments at the bottom of lakes. These may date back thousands of years and are sometimes laid down yearly in layers called varves, like the rings built up yearly by growing trees. The sediments preserve the remains of plankton from the water and pollen from nearby forests that may no longer exist. The records allow scientists to reconstruct the history of the area. They can also record onetime events, such as ash from an ancient volcano or recent radioactive pollution.

Lakes in Sweden usually freeze over in winter. Birds cannot reach food such as fish and water plants. Most animals beneath the ice decrease activity until spring.

small creatures that sift through the mud for detritus (pieces of dead material). If the river brings enough mud and there is no powerful sea current to take it away, it may build its own flatland (a delta) far out at sea. Lagoons, brackish (semisalty) marshes, and sometimes mangrove swamps also contribute to the natural variety at the river's end.

How Lakes Work

Compared with rivers and streams, lakes (especially large ones) depend more on food produced in the water itself than on dead material falling in from the shore. As in the sea, the food chains depend mainly on microscopic floating algae, or phytoplankton. Phytoplankton grow only in the upper layer of the lake, where there is enough light. They need chemicals called nutrients, which are dissolved in the water. The nutrients determine how many phytoplankton can survive. Scotland's Loch ("lake") Ness, for example, is very low in nutrients, so there are few phytoplankton. It would be very difficult for Loch Ness's legendary monster to find enough to eat in such infertile water.

The way the waters of a lake mix—or don't mix—greatly affects the lake's life. The sun warms the surface water. Because warm water is less dense than cold water, it floats on top as a separate layer. This top layer has plenty of oxygen because the waves mix with the air. This is good for the animals that live there but as they grow, they use up nutrients, and when they die, they sink to the layer below. The upper layer can therefore be very low in nutrients. The layer below is just the opposite: It has less opportunity to receive oxygen and can't support very much life, but it has plenty of nutrients.

In cooler countries, the warm surface layer forms in summer but disappears in fall, and winter winds can mix the lake waters completely. If ice forms, the waters mix a second time in spring, when the ice melts. Thanks to the frequent mixing, oxygen is available deep down, allowing burrowing animals such as worms to survive in the mud.

In tropical lakes, the layers may be almost permanent. The deep water is not only cooler but has far less oxygen—sometimes so little that only microorganisms can survive. When dead matter sinks to the lake floor, the microorganisms break it down and release poisons such as hydrogen sulfide. On the rare occasions when such tropical lake water mixes, millions of fish may die as the rising stagnant water suffocates and poisons them.

Near the edge of a lake, the water may be sufficiently shallow and well-lit for reeds and other plants to take root in the muddy bottom. The shallows are often ideal habitats for insects, snails, and other creatures, and the plants provide hiding places for fish. In other parts of the lakeshore, though, waves and wind may prevent plants from growing, making these rocky shores look barren.

Salt Lakes

Although most lakes contain freshwater, some contain water as salty as the sea—or even saltier. The world's largest salt lake, the Caspian Sea, was once joined to the ocean. Today, it still contains mainly marine organisms.

Most salt lakes lie far from the sea, however, and are salty because, like the Caspian Sea, they have no outlet. Rivers that drain into lakes always bring some dissolved salts with them, so a lake only stays fresh if it also has an outlet, allowing the salts to wash away. Otherwise the salt has nowhere to go and gradually builds up. Sometimes the main salt is common salt (sodium chloride), but other lakes contain mainly soda (sodium carbonate), which makes the lake alkaline. (An alkali is the opposite of an acid.) Most fish cannot survive in very salty or alkaline water, but certain algae, shrimps, and other animals thrive in such habitats. With little competition, they multiply in lakes such as Utah's Great Salt Lake. In East Africa's soda lakes, algae and shrimps feed millions of flamingos (below).

Great Lakes

The Great Lakes of North America hold nearly one-fifth of Earth's surface freshwater. The hollows now occupied by the lakes were carved out during several ice ages.

Isle Royale, the land stretching across the top of the picture, lies in Lake Superior and is home to timber wolves and moose. It became a world biosphere reserve in 1980.

1. Boundary Waters

Thousands of smaller lakes on the Minnesota–Ontario border draw canoeists in search of wilderness adventures. Few anglers reach the more remote lakes, where smallmouth bass, northern pike, walleye, and lake trout are all abundant.

2. Isle Royale National Park

The islands of this park are surrounded by shipwrecks now explored by divers.

3. Keweenaw National Historical Park

Copper has been extracted here by Native Americans since 7,000 years ago. The park also preserves relics of what became the world's most important source of copper in the 19th century.

4. Lake Superior

The world's largest freshwater lake by area, Lake Superior is the size of South Carolina. The lake stores more water than all the other Great Lakes put together. It is called "Superior" because it is the highest Great Lake above sea level. At up to 1,333 feet (407 m) deep, it is also the deepest.

5. Lake Michigan

With a coastline ranging from marshes and prairies to sand dunes, Lake Michigan's ecology is varied. Chicago and other cities lie on its shores. It is nearly as large as West Virginia.

6. Sleeping Bear Dunes National Lakeshore

This region of lakeside dunes towers up to 460 feet (140 m) above Lake Michigan's waters.

7. Fathom Five National Marine Park

This underwater park preserves evidence of the Great Lakes' history, including a huge former waterfall that once allowed Lake Huron's waters to reach the sea via the Ottawa River.

 ## Fact File

▲ The Great Lakes store and absorb so much heat that they cause winters to be milder and summers cooler than in prairie lands farther west.

▲ Much of the lakes' surfaces freeze over for four or five months in winter.

▲ The easy transportation of grain, coal, copper, and other commodities via the Great Lakes was a major factor in the industrial and agricultural development of the United States.

NORTH AMERICA

Right: The American Falls at Niagara do not have the spectacular horseshoe shape of Niagara's Canadian Falls, but at 167 feet (51 m), they are a little taller.

0 — 200 miles
0 — 200 km

N

Hudson Bay   QUEBEC

12

CANADA

Lake Nipigon

Ouimet Canyon

Pukaskwa National Park

ONTARIO

4

Lake Superior

Lake Superior National Park

Quetico Provincial Park

Voyageurs National Park

Thunder Bay

2

Boundary Waters

1

Isle Royale National Park (U.S.)

3

Whitefish Bay

MICHIGAN

Apostle Islands National Lakeshore

Keweenaw National Historical Park

Green Bay

Minneapolis

WISCONSIN

5

6

Sleeping Bear Dunes

North Channel

Manitoulin Is.

Fathom Five National Marine Park

7

Bruce Peninsula National Park

Killarney National Park

Georgian Bay

Georgian Bay Island National Park

Ottawa River

Ottawa

Haliburton Highlands

St. Lawrence Islands National Park

St. Lawrence River

Niagara Escarpment

8

Lake Huron

Mississippi River

MINNESOTA

Milwaukee

Lake Michigan

Toronto

Lake Ontario

13

IOWA

ILLINOIS

Chicago

Indiana Dunes

MICHIGAN

Lake St. Clair

Detroit

Point Pelee National Park

9

ONTARIO

10 Lake Erie

Mayfly plague area

12

Buffalo

Niagara Falls

NEW YORK

11

PENNSYLVANIA

UNITED STATES

OHIO

Cleveland

8. Lake Huron
This lake has around 90,000 islands and is almost cut in two by a ridge of hard rock called the Niagara Escarpment, which also forms Niagara Falls.

9. Point Pelee National Park
Canada's southernmost point is a major center for migrating birds, as well as containing broad-leaved forest that was once widespread in the region.

10. Lake Erie
The shallowest of the Great Lakes, Lake Erie is bordered by cities such as Detroit and Cleveland. Pollution increased after 1950 and led to one of the lake's inlets catching fire due to all the oily pollutants it carried. A campaign to clean up the lake was launched.

11. Mayfly Plague
Lake Erie has been cleaned so successfully that the larvae (young forms) of mayflies are again abundant in the lake. In summer, the adults emerge from the lake and swarm in huge numbers. They quickly mate, lay eggs, and die, their bodies piling up in heaps.

12. Niagara Falls
These famous falls are slowly moving upstream as the water erodes the underlying rocks. The nearby Welland Canal allows ships to bypass the falls and reach Lake Erie.

13. Lake Ontario
Lying below Niagara Falls, Lake Ontario has always been more accessible to fish and other animals that reach it from the sea via the St. Lawrence River.

Changing Ecology

At the end of the last ice age, fish colonized the newly ice-free Great Lakes, while trees took over most of the nearby land. When Europeans began to settle in the 1700s, they cut down many of the forests, choking tributary streams with sawdust so fish such as trout and sturgeon could no longer lay eggs there. The lakes provided commercial fishers with large catches of trout and plankton-eating fish such as whitefish and lake herring, but overfishing meant that by the middle of the 20th century these catches were declining. The Welland Canal also allowed in alien species such as the sea lamprey (below), an invader that eats the native fish alive.

Water Plants

From microscopic plankton to towering palm trees, many plants or plantlike organisms make their home in rivers and lakes. Water is seldom in short supply, but getting enough light, air, and nutrients can be a challenge.

Plants growing in water face a different world from their land-living relatives. One crucial challenge they face is getting oxygen. Water plants need to breathe air, just as we do, but they often grow in mud that contains none of this vital element at all. Many of them make up for this with air-filled stems and leafstalks. These transport oxygen down to their roots—and often help keep the plants from sinking in the water as well.

Plants need light, but light can be in short supply underwater. In streams, there's also the constant danger of being washed away by the current. On the other hand, the biggest challenge to land plants—getting enough water—is much less of a problem for aquatic plants. In fact, most water plants can take in water through their leaves and stems, rather than relying only on their roots. If a pond does dry out, though, some species become temporary land plants, growing toughened leaves that resist wilting in air.

Small and Simple

Among the most important plants in freshwater are algae—a name given to a broad range of organisms, from microscopic single cells to giant seaweeds. Strictly speaking, algae are not true plants. However, like true plants, they use the sun's energy to synthesize their own food, a chemical process called photosynthesis.

Most freshwater algae are the microscopic, single-celled variety. In lakes particularly, these play a crucial role in the plankton—the community of tiny, free-floating organisms that forms the basis of food chains in both the sea and freshwater. Planktonic algae are also called phytoplankton (plantlike plankton) to distinguish them from zooplankton (animal-like plankton). Other types of algae grow attached to rocks or the surface of larger plants, where they are sometimes visible as a fuzzy green layer.

Insects such as this honeybee find floating water plants handy as a secure platform on which to drink.

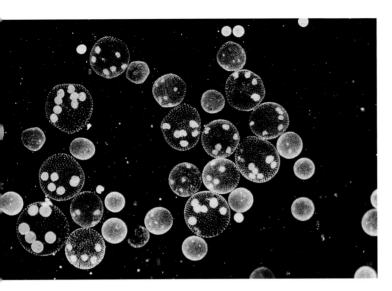

Microscopically tiny algae called volvox form ball-shaped colonies less than 0.04 inch (1 mm) across. They float in the water like little green jewels.

Freshwater phytoplankon are very diverse. The smallest are cyanobacteria, which sometimes multiply until they form a thick green scum on the water. Cyanobacteria were once called blue-green algae, but they are neither true plants nor true algae. Instead, they are considered to be types of bacteria (though very different from the bacteria that cause human diseases). Other microscopic algae, called diatoms, are noted for making beautiful, glasslike shells out of silica, the same substance that sand is made of.

In many lakes, there is a spring bloom of phytoplankton growth, fueled by increasing light and warmth. When phytoplankton use up the lake's nutrients, the bloom finishes and the remains of dead phytoplankton fall to the lake bed.

Larger Plants

Most large freshwater plants are flowering plants, which have dominated plant life since they first evolved more than 100 million years ago. In a few freshwater habitats, such as mountain streams, moss, ferns and other nonflowering species can also be common.

Freshwater flowering plants evolved from various land-living families. They are easier to describe in terms of the way they grow rather than in terms of the families of plants they belong to. For example, plants such as reeds, which grow through shallow water and into the air, are known as emergents. In deeper water, these are usually replaced by plants such as water lilies, which have floating leaves but roots that are anchored in the bed. Other plants float freely on the surface with dangling roots, and some live entirely underwater. Finally, some plants are adapted to life in fast streams, either clinging tightly to rocks or growing long, flexible stems that wave in the current.

Reeds, cattails, and other emergents can form dense stands along the edge of a river or lake. They spread quickly, sending horizontal stems (rhizomes) through the mud. At intervals, the stems produce new shoots—the parts you see above water. On the edge of a pond, the dead remains of emergent plants can build up to form a waterlogged soil called peat. Peat is very fibrous and microorganisms can't break it down due to the lack of oxygen. Peat builds along the pond edges, and the pond may gradually fill up. Trees and shrubs may take over, forming a swamp or even dry land.

New to Science

In 1992, plant experts working on the island of Madagascar in the Indian Ocean made an exciting discovery: They found the world's only palm tree that begins life underwater. Now known as the water palm, this tree grows only in a single river system on the island. Its fruits fall into the water and float, later splitting to release a seed that has already started to sprout. This sinks to the sandy river bottom, where it takes root and grows to maturity.

Common reeds are emergent water plants, rooting themselves in water shallow enough for their stems to reach high above the water and breathe in the air.

Water lilies, probably the best known water plants, also grow using rhizomes. The rhizomes store food and are eaten by people in some parts of the world. The biggest species—the giant Victoria lily of the Amazon—has round, tray-shaped leaves up to 6.5 feet (2 m) wide.

Water lilies breathe through tiny airholes on the upper surface of their leaves; the leaves produce wax, which prevents the leaves from filling with water. Other plants that usually have floating leaves include water crowfoots (relatives of the buttercups), duckweeds, and pondweeds, which are often important food for ducks.

Many mosses cling to stones around the damp edges of mountain streams and waterfalls, although a few live their whole lives underwater. Mosses are important to small animals because they provide shelter from the current.

Around tropical waterfalls and rapids you can also find some very weird flowering plants that glue themselves flat to rocks to avoid being swept away. Known as podostemons, they look like green chewing gum—except in the dry season, when the "chewing gum" starts sprouting flowers.

Right: Mosses flourish on the water-splashed rocks beside many mountain streams and provide shelter for a host of tiny aquatic creatures.

 Sacred Flower

The sacred lotus of Asia is a water plant with large emergent leaves and flowers like those of a water lily. It is regarded as sacred in the Buddhist religion and has appeared in countless religious drawings and carvings. There is also a lotus flower native to North America. Some botanists regard this as a subspecies of the sacred lotus, while others consider it a separate species.

The duckweed family includes Earth's smallest flowering plants. Valdivia duckweed (above) is one of the tiniest. The rice grain in the center shows just how small it is.

Floaters and Sinkers

Plants that float freely in the water risk being swept away, so they thrive best in still waters. They also need nutrient-rich water, because their roots are not in contact with soil or mud. Under the right conditions, however, they can entirely cover the surface of ponds and ditches, smothering other plants and preventing oxygen from reaching the water below. Floating plants include many species that are now invasive pests, such as water hyacinth, water lettuce, and several floating ferns. They also include the miniature duckweeds. The whole body of a duckweed is a few tiny, floating oval leaves bearing a few small roots. *Wolffia* duckweed has leaves less than 0.04 inch (1 mm) long.

The most specialized water plants are those that live completely submerged. They often grow in shady places and must absorb all their oxygen and other gases from the water. Often these submerged plants have finely divided leaves that help them absorb gases and nutrients. Among the most widespread are the milfoils (a name coming from the French for "a thousand leaves"), including the native American northern milfoil and its invasive Eurasian cousin, the spiked milfoil.

Floating in Ambush

Bladderwort is the only carnivorous (animal-eating) plant that lives underwater. Floating just below the surface in ponds, it grows small swellings called bladders. If an animal, such as a mosquito larva (left) or a water flea, bumps into a bladder, it is sucked in, trapped, and later digested, providing the plant with useful nutrients such as nitrogen. Although there are other carnivorous plants, they all grow in bogs or fens rather than in open water. They all trap insects to get nutrients not present in the soil.

Water crowfoots have some of the prettiest of all water flowers. They are like buttercups with floating leaves and flowers, and they bloom in clean freshwater.

The Next Generation

Like plants on land, aquatic flowering plants can reproduce either sexually—via flowers and seeds—or asexually. Asexual reproduction can be accomplished by creeping stems or roots, or it can simply rely on small fragments of the plant breaking off and taking root elsewhere. Such asexual methods are particularly common in water plants, probably because it's easy for plant fragments to float away to new habitats without drying out on the way.

Besides helping plants spread, asexual reproduction can help them survive being frozen and broken apart in winter. Many water plants produce specialized winter buds called turions, which can survive on the bottom of a pond below the surface ice while the rest of the plant dies. In the case of a common floating plant called water soldier, the whole plant sinks to the bottom before sprouting smaller plants at the end of short stems. In spring, they return to the surface, where they give rise to a new generation.

Most water plants, even submerged ones, produce their flowers above water. They rely on wind or insects to transfer the pollen (which contains the male sex cells) from one flower to another, a process called pollination. When the female part of a flower is pollinated, a seed develops within it. Plants such as reeds and cattails are wind-pollinated; their small flowers produce dusty pollen that can blow for miles in the wind. Water lilies and other species with showy flowers rely on insects to pick up sticky pollen. They attract the insects with scent, colorful petals, and a sugary secretion called nectar. Some water lilies even generate heat as an attraction.

A few water plants, including one called hornwort or coontail, have gone one step further: They flower underwater. Such plants produce threadlike pollen that can wrap around the female parts of flowers. Another plant—water celery—produces male flowers that detach and float to the surface like tiny boats. These drift until they reach a female flower (which remains anchored), allowing pollination to take place.

After pollination, the flowers of aquatic plants may withdraw underwater to protect the growing seeds. The stalks of water lily

Water milfoils can live completely submerged. The pictured species is native to South America, but it has escaped from aquariums in the United States.

Water lilies can thrive even in deep water, anchoring themselves to the water bed by underwater stems often more than 10 feet (3 m) long.

flowers twist into a corkscrew shape to accomplish this. The fruits and seeds often float, helping to disperse them.

Plant Invaders

Some of the worst weeds in the world are water plants. People have unintentionally helped these weeds spread far beyond their original ranges and invade foreign countries. Some have been grown as ornamental plants in garden ponds and then escaped into the wild. Others have been sold around the world as aquarium plants; when people empty their aquariums into lakes or ditches, the plants escape. In the new habitats, there may be no animals capable of eating the plants, with the result that they start multiplying out of control. Once a foreign species becomes established, boat trailers often help it along by transporting it from lake to lake.

In the United States, southern states are at risk from tropical and subtropical weeds, while farther north, several alien plants from Europe and Asia have taken hold. Even native species can become weeds—for example, the fragrant water lily, an attractive plant native to eastern North America, has now invaded lakes much farther west.

One notorious weed of warmer climates is water hyacinth, a floating plant from South America that has large, glossy leaves and striking purplish flowers. It multiplies asexually by producing horizontal underwater stems that give rise to new plants. In a short time, water hyacinth can spread to cover a large area of water with a dense mat of plants, preventing oxygen from reaching the water and choking the native wildlife. A whole scientific journal is devoted to this one plant and the problems it causes.

Destroying waterweeds is now a huge industry. The battle against the invading plants involves several different lines of attack, including mechanically removing plants from the water; poisoning them with herbicides (plant-killing chemicals); biological control (introducing organisms such as insects that eat the plants or cause disease); and educating people to be careful when they move boat trailers.

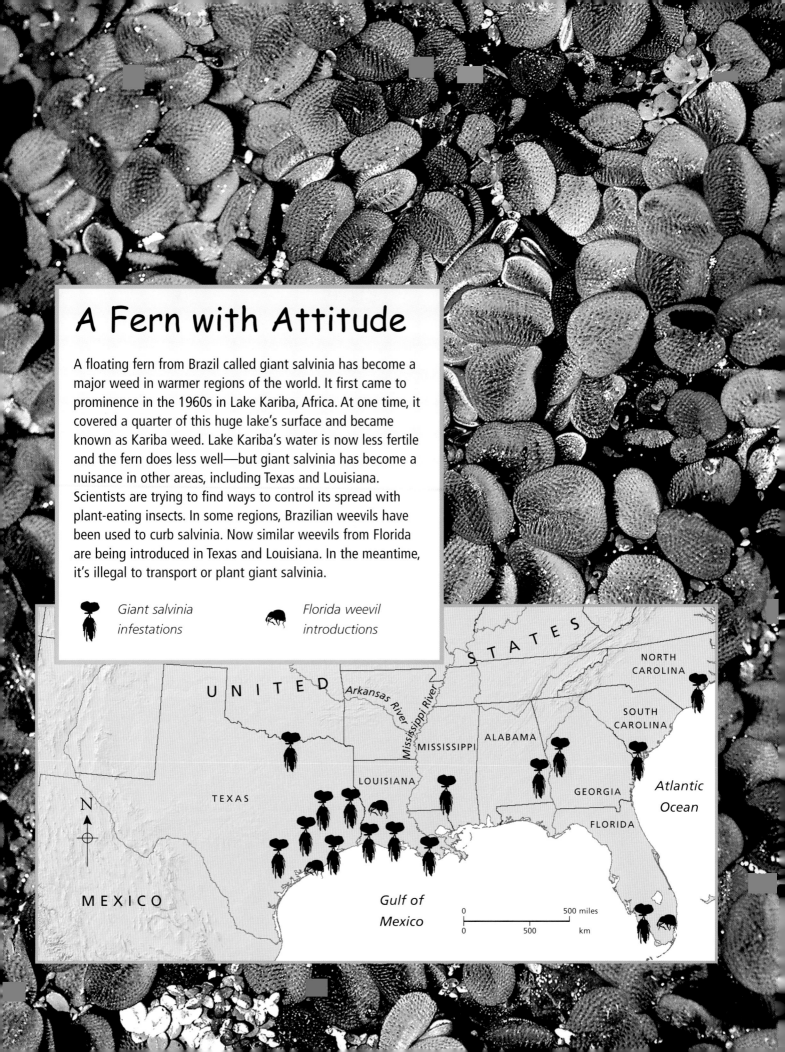

A Fern with Attitude

A floating fern from Brazil called giant salvinia has become a major weed in warmer regions of the world. It first came to prominence in the 1960s in Lake Kariba, Africa. At one time, it covered a quarter of this huge lake's surface and became known as Kariba weed. Lake Kariba's water is now less fertile and the fern does less well—but giant salvinia has become a nuisance in other areas, including Texas and Louisiana. Scientists are trying to find ways to control its spread with plant-eating insects. In some regions, Brazilian weevils have been used to curb salvinia. Now similar weevils from Florida are being introduced in Texas and Louisiana. In the meantime, it's illegal to transport or plant giant salvinia.

Giant salvinia infestations

Florida weevil introductions

Amazon

The sheer scale of the Amazon River is awe inspiring. One-fifth of all the world's river water passes between its banks.

Even small tributaries in the Amazon river system are major rivers (below). They provide a system of watery highways, useful for travel in a region with few roads.

Fact File

▲ The Nile may be longer than the Amazon, but the Amazon is definitely Earth's greatest river—it drains the largest river basin in the world.

▲ The Amazon has far more types of fish than any other river system, with between 2,000 and 3,000 different species living within the basin.

▲ The Amazon provides some unusual habitats: Plants float on the river, forming islands that may be more than a half mile (0.8 km) long. They sink and decay during the dry season.

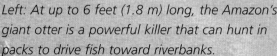

Left: At up to 6 feet (1.8 m) long, the Amazon's giant otter is a powerful killer that can hunt in packs to drive fish toward riverbanks.

28

Map labels:
- Llanos (grassland)
- Orinoco River
- VENEZUELA
- Mount Roraima ▲
- Georgetown
- Paramaribo
- GUYANA
- Cayenne
- COLOMBIA
- Guiana Highlands
- Angel Falls
- Iwokrama reserve
- SURINAME
- FRENCH GUIANA
- Boa Vista
- **2** →
- Casiquiare Channel
- Pico da Neblina ▲
- Amazon
- **8**
- Atlantic Ocean
- N
- 500 miles
- 500 km
- **1**
- **11** Mouths of the Amazon
- Equator
- ECUADOR
- **1**
- Putumayo River
- Napo River
- Rio Negro (Negro River)
- Branco River
- **4**
- Manaus
- Amazon River
- Tidal marshes
- Marajó Island
- Belém
- Iquitos
- Rio Abiseo National Park
- Japurá River
- Jaú National Park
- **10**
- Tucuruí Reservoir
- Flooded forest
- **3**
- **5**
- Madeira River
- **7**
- Trans-Amazon Highway
- Tapajós River
- Iriri River
- Xingu River
- to Recife
- Trujillo
- Juruá River
- Flooded forest
- Cruzeiro do Sul
- Purus River
- **6**
- **9**
- rain forest
- **9**
- Ucayali River
- **1**
- Andes
- PERU
- B R A Z I L
- Araguaia River
- Tocantins River
- Brazilian Highlands
- **9**
- Pacific Ocean
- Lima
- Manu National Park
- **1**
- BOLIVIA
- Paraguay River wetlands 280 miles (450 km) south ↓
- Brasilia
- SOUTH AMERICA

1. Amazon Headwaters
Many Amazon tributaries begin up in the Andes of Bolivia, Peru, Ecuador, and Colombia. Even in minor tributaries there are many fish species, including tetras, bony tongues, catfish, piranhas, and electric eels.

2. Casiquiare Channel
A natural river connection between the Rio Negro and the Orinoco River to the north. Such a connection between two watersheds is very unusual. It is 300 feet (500 m) or more wide. In flood, it flows at 5 mph (8 km/h) toward the Rio Negro.

3. Flooded Forest
The Amazon and many of its tributaries flood every year. In the flooded areas, trees grow that can tolerate being in up to 40 feet (12 m) of water for several months of the year.

4. Rio Negro
This black-water tributary is Earth's second largest river in terms of volume of water flow.

5. Madeira River
The valley of this tributary is a flyway for waterbirds migrating between the Amazon and the Paraguay River wetlands.

6. Amazon Rain Forest
The Amazon drains the greatest rain forest in the world. Despite rapid deforestation, much of this forest is still intact.

7. Trans-Amazon Highway
This road has opened much of the previously remote Amazon region, with effects that include leakages of poisonous mercury into rivers from gold mining.

8. Iwokrama Reserve
Part of the effort to protect Earth's largest freshwater fish—the endangered 13-foot (4.5-m) arapaima.

9. Clear-water Rivers
A diverse range of fish migrate in vast numbers up and down clear rivers such as the Xingu, Araguaia, and Tapajós.

10. Tidal Marshes
Ocean tides push back the Amazon's waters daily. They create tidal forests and marshes that are flooded with river water twice a day up to 260 miles (400 km) inland.

11. Mouths of the Amazon
The strength of the Amazon's current carries its waters up to 200 miles (320 km) out to sea.

Three Kinds of River

The rivers of the Amazon Basin are of three main kinds: white water, clear water, and black water. White-water rivers include the Amazon itself and its major southern tributary, the Madeira. White-water rivers, which are really light brown, gain their color from the sediment they carry down from the Andes mountains. Fish that live in these rivers use sound, smell, and touch—it is difficult to see far in the murky waters. Clear-water rivers drain the much more ancient, worn-down rocks of the Brazilian Highlands, so they carry little sediment. Black-water rivers also carry little sediment, but dissolved substances from decaying plants give the water a dark color. Black-water rivers include the Rio Negro (right), which meets the white-water Amazon near Manaus.

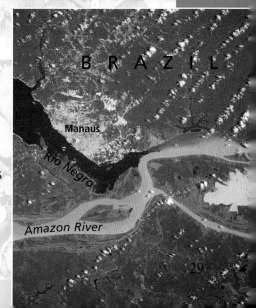

BRAZIL
Manaus
Rio Negro
Amazon River

Water Animals

Freshwater habitats contain an amazing variety of animals. The majority are invaders—animals whose ancestors entered freshwater from either the ocean or from land.

For sea-living animals, making the move to freshwater has its challenges. Freshwater habitats are more changeable than the sea—they may freeze, turn into raging torrents, lose all their oxygen, or even dry up altogether. There are also problems caused by the lack of salt. While seawater has about the same salt concentration as the fluid in animals' bodies, freshwater is much less salty. The salty fluid in an animal's body tends to absorb freshwater, making the body fluid too watery. Small animals are always in danger of water flooding into their tissues, making them swell up and even burst.

As a result of these challenges, not all sea animals have made the leap to living in lakes and rivers. For example, while there are freshwater crabs in many parts of the world, you don't find freshwater starfish or freshwater octopuses.

Water animals whose ancestors lived on land include insects and most freshwater snails, in addition to mammals such as otters. To live in freshwater, all such animals need

to return to the water surface to breathe air, unless during millions of years of evolution in freshwater, they have developed gills. Gills are feathery organs divided into a myriad of tiny branches. They have an enormous surface area, and they allow dissolved gases—oxygen and carbon dioxide—to pass easily between an animal's body and surrounding water. The gas exchange is the equivalent of breathing underwater. No mammals have developed gills, but gills are common among freshwater insects, especially in their underwater immature stages (larvae).

Of all the animals to have colonized freshwaters, fish and insects are probably the most successful. Some 10,000 species of fish of all kinds live in freshwater. Insect types that have made freshwaters their own include dragonflies, mayflies, mosquitoes and midges, caddis flies, bugs, and beetles. On the left, a brown trout leaps to catch a damselfly.

Left: Larvae of insects can be both prey and
predator in freshwater food chains. In ponds, diving
beetle larvae called water tigers (on the right) prey
on hapless mosquito larvae (on the left).

Many freshwater invertebrates, like their marine relatives, live by filter feeding—they strain tiny particles of food from the water. Freshwater sponges and mussels get their food this way. Another feeding method, used by many worms, is to take mouthfuls of mud and extract food particles from it. Some invertebrates are parasites—the leeches that can suck your blood are one example—while many snails and insect larvae feed by scraping algae off rocks and plants. Larger creatures such as crabs and crayfish take bigger food items, sometimes virtually anything they can get hold of, living or dead.

These varied creatures have to deal with the uncertainties of freshwater life. A trick that many of them share is the ability to produce eggs that resist drying out, or even to seal up their whole bodies in temporary protective cocoons. If a pond or stream dries up, not only will many of these small animals

Small and Spineless

When classifying the living world, biologists divide animals into major groups. One group is the vertebrates (animals with backbones). All the other groups are invertebrates (animals without backbones), but there are many unrelated types. Most small animals, such as worms and insects, are invertebrates.

Invertebrates are a vital part of all water ecosystems. Some types are better known from the sea, such as sponges and crabs. Arthropods (invertebrates with jointed legs) are particularly common in freshwater. They include insects, spiders, water mites (minute relatives of spiders), and crustaceans (crabs, crayfish, and shrimps). Freshwater snails and clams are also common.

Zooplankton

Zooplankton form an important part of some freshwater ecosystems, especially those of large lakes. They are tiny animals and animal-like microorganisms that float freely as part of the plankton. In the sea, many zooplankton are larvae of larger creatures such as clams, but in freshwater, such larvae are much less common. Instead, common freshwater zooplankton include water fleas (right), which are crustaceans, not related to real fleas, and a unique group of microscopic animals called rotifers. Most zooplankton eat plantlike plankton (phytoplankton), but some are carnivores (meat eaters), eating smaller zooplankton. In turn, zooplankton provide food for fish and other larger animals.

The feathery red gills of this axolotl (a type of salamander) help it take the oxygen it needs from the water and remove carbon dioxide from its body.

survive, but they may be carried by the wind, or in the guts of larger animals, and transported elsewhere. Many small freshwater animals have a continental or worldwide distribution, most likely for this reason.

Obtaining oxygen is also often a problem. Animals need this gas so their bodies can release energy from food, a process called respiration. Though air contains plenty of oxygen, water contains less, especially if it is very still or stagnant. Some freshwater snails rely on gills to extract oxygen from the water, but they can only live in water that is high in oxygen. However, other snails have lungs, so they can come up for air if they live in stagnant water. Bloodworms (relatives of earthworms) manage to live in oxygen-poor mud, using the red pigment in blood (hemoglobin) to store oxygen.

Salt lakes are very harsh environments, but one type of crustacean—the brine shrimp—thrives in them. A few other crustaceans can

even live in hot springs at temperatures of up to 130°F (55°C). The salty mud of estuaries (where rivers meet the sea) is also rich in invertebrates, mainly marine types that can tolerate the mixing of salt and freshwater.

Insect Invasion

There are more kinds of insects in the world than all other animal species put together. Most live entirely on land, but over millions of years several major groups have become freshwater specialists, and many habitats would be difficult to imagine without them.

It is only the larvae of many freshwater insects—including mayflies, caddis flies, dragonflies, and mosquitoes—that live underwater, while the adults live in the air. In the case of water beetles and water bugs, however, the adults live in water, too.

Insects play many different roles in freshwater ecosystems. Some are filter feeders, while others specialize in eating dead leaves

Masters of the Air

Dragonflies are among the most eye-catching of all insects. These powerful, large-eyed flyers sometimes migrate hundreds of miles and are so aerobatic that they can even fly backward. Despite their long, needle-like bodies, they have no sting; however, they are fierce predators and can snatch other flying insects in midair. They lay their eggs underwater (below). The eggs hatch to begin the underwater stage of their life cycle.

Along with their more delicate relatives, damselflies, dragonflies spend the bulk of their lives as underwater larvae, or nymphs. A dragonfly nymph is just as fierce a predator as an adult. Lying in wait among stones or plants, nymphs snatch passing prey using a unique extendible device. This hinged structure is normally folded beneath the head until the nymph shoots it out suddenly, seizing an unsuspecting victim.

egg-laying tube

or in scraping algae and other matter from rocks and plants. Many water beetles and water bugs are predators. Water beetles use their mouthparts to catch and eat prey, while water bugs have piercing-sucking mouthparts to drain their victims dry. Some of the giant water bugs, also called electric light bugs (because the adults often fly toward lights), are among the largest of all insects—so big that they can catch and eat small frogs.

Aquatic (water-living) insects have evolved many features that equip them for life in the water. Some have snorkels for breathing through the surface, or gills for extracting oxygen from the water; others carry air bubbles around with them. Insects in rushing streams, such as many mayfly larvae, often have flat bodies that help them avoid the current and wriggle under rocks. Some caddis fly larvae build themselves

Trout fry (young) hatch in rushing mountain streams. They burrow into the gravel and live on their yolk sacs until they grow big enough to swim against the current.

protective cases out of small stones or debris, and others weave underwater nets to catch prey. Some predatory bugs, such as water striders, literally walk on water, using their sensitive feet to feel for vibrations made by flies trapped in the water surface.

The ability to fly provides insects with an advantage in spreading to new habitats, and most freshwater species, such as water beetles, have held onto this ability. Flying is also useful in finding a mate—male mayflies and midges, for example, each form large, dancing swarms that attract the females.

Freshwater Fish

Worldwide, there are about 25,000 fish species—as many as all the other vertebrates (birds, mammals, reptiles, and amphibians) put together. Of these, about 10,000 live in the world's rivers and lakes—even though freshwaters are tiny in area compared to the oceans. This high proportion is probably due to lakes and rivers being separated by land, allowing isolated populations to evolve into new species. In the lakes of Africa, there are hundreds of different fish species all belonging to the cichlid family. These evolved in isolation after the lakes formed.

Because of their size, speed, and sharp senses, fish dominate many rivers and lakes. They are often the top predators in the water—though many fall prey to land animals, such as birds and bears. The feeding habits of fish vary widely. Some lake fish feed by straining tiny plankton through their gills, but most are predators, eating invertebrates,

Freshwater crabs protect their gills inside a body chamber so they can keep them moist during forays over land.

other fish, or sometimes even waterbirds and mammals. The fearsome red-bellied piranha of the Amazon appears to hunt in groups, ganging up on unsuspecting victims before attacking them with razor-sharp teeth. Some fish, including several African cichlids, specialize in tearing only the scales off other fish. A few are plant eaters. One species, the grass carp, eats underwater plants; people deliberately release them in rivers in the United States to help get rid of waterweeds. Sterile grass carp are used so they don't breed and become a problem themselves.

Many kinds of piranhas lurk in the Amazon river system. Some are aggressive predators, but others eat only berries dropped by trees in the flooded Amazon forest.

Varied Lifestyles

Freshwater fish have evolved in varied ways, reflecting their diverse lifestyles. Species such as trout, which live in fast-flowing streams, need to swim constantly and usually prefer cool water, which provides more oxygen. They are streamlined, with tail fins shaped for rapid acceleration toward prey or away from danger. Other stream-living fish use their mouths as suckers to anchor themselves against the current. Fast streams are also liable to carry away eggs, but fish overcome this either by making nests in gravel, laying sticky eggs, brooding eggs in the mouth, or giving birth to swimming young. A fish called the splash tetra even leaps out of the water to lay its eggs on overhanging leaves.

Kinds of Freshwater Fish

Rivers and lakes shelter some of the most ancient families of fish on Earth, besides many more recently evolved kinds. The oldest of all are lampreys (below), which don't have jaws but feed by attaching to fish and rasping at their flesh with suckerlike mouths.

Most fish have jaws, though, and belong to two main groups: bony fish, which have skeletons made of bone; and cartilaginous jawed fish (sharks and rays), which have skeletons made of cartilage (gristle). Sharks live mainly in oceans, but some can cope with freshwater. Bull sharks—one of the few human-eating species— swim far up major rivers including the Mississippi.

The great majority of freshwater species are bony fish. Among these, the six species of lungfish hold a unique place in evolution because ancient relatives of these fish are thought to be the ancestors of all land-living vertebrates, including ourselves. Other ancient groups still alive today include bowfins, paddlefish, and garpikes of North American waters, and the prehistoric-looking sturgeons. Some scientists suspect that bony fish first evolved from cartilaginous fish in freshwater, then gradually invaded the ocean. That might explain why members of so many prehistoric fish families live in freshwater today.

Fish that feed on the muddy bottoms of rivers, such as catfish, sturgeons, and rays, tend to be flat with downward-facing mouths; often they are also well camouflaged. The unique four-eyed fish, or anableps, hunts at the surface for its food. Each of its two eyes is divided into two halves, one half for seeing in air and the other for seeing in water.

Compared with the sea, freshwaters are often muddy, with poor visibility, but freshwater fish have a superb array of sensory organs that help them get around this. Like all fish, they have a vibration-detecting system called a lateral line running down each side of the body. The lateral line can feel the ripples made by other moving

 Electric Fish

Electricity plays an important part in the lives of many freshwater fish—notably the electric eel (right) of South America and the electric catfish of Africa. Both these animals can produce electric shocks of several hundred volts—enough to stun both prey and potential predators. Many other freshwater fish produce weak electric discharges, often as rapid pulses. Combined with electrical detectors in the skin, these discharges are used like radar to detect nearby objects in dark or muddy water, as well as for signaling between individuals of the same species. Many fish that do not generate electricity can nevertheless detect it. Some predatory fish, including freshwater stingrays, are sensitive to the tiny electric discharges made by their victims' muscles, for instance.

Although it is eel-shaped, the electric eel is not a member of the eel family and is not a true eel. The dimples in its skin are electrical sense organs.

objects, giving a sense of touch at a distance. Most fish also have a good sense of smell; salmon can detect the faint but distinctive scent of the river in which they were born, allowing them to return to it after years spent at sea. Catfish, which hunt mainly at night, usually have poor vision but make heavy use of the chemical sensors on the barbels around their mouth; some catfish can even taste with their tail. Many fish also use electricity to scan their surroundings.

Sturgeons have been around since the Jurassic period, 150 million years ago.

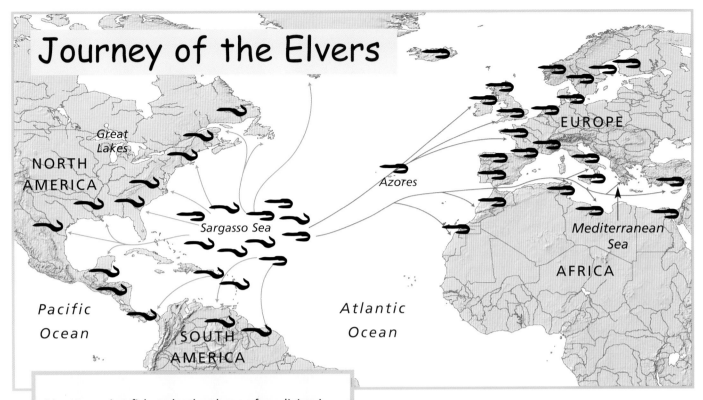

Journey of the Elvers

Great Lakes

NORTH AMERICA

EUROPE

Azores

Sargasso Sea

Mediterranean Sea

AFRICA

Pacific Ocean

SOUTH AMERICA

Atlantic Ocean

Many amazing fish make the change from living in seawater to living in freshwater when they migrate. But whereas salmon and sturgeon swim up rivers from the ocean to lay eggs, both the European eel and the American eel use exactly the opposite migration pattern. Their adult life is spent in rivers and lakes in the Americas, Africa, and Europe, but they return to the depths of the Sargasso Sea to breed (and to die). The next generation of young eels (elvers) must then find their long way back to freshwater. Only 2–4 inches (5–10 cm) in length, the elvers swim hundreds of miles to reach rivers, where they feed and grow.

→ European eel

→ American eel

Fish normally use gills to extract oxygen from water, but the oxygen supply in freshwater can be unreliable, so various freshwater fish can also breathe air. Some catfish swallow air, for example, and eels can breathe through their skin. Electric eels absorb oxygen through the lining of the mouth—they drown if kept underwater permanently. Lungfish, as the name implies, have lungs. They can survive buried in mud even if their pond or lake dries up.

Some fish migrate long distances as part of their life cycle. Salmon spend their adult lives in the sea but swim up rivers to breed in mountain streams, perhaps because their

Turtles are reptiles, but like amphibians, many live partly on land and partly in freshwater. Unlike amphibians, though, they lay their eggs on land. Reptile embryos would drown inside the egg if they were laid underwater. This painted turtle lives in the Mississippi.

All herons are expert fishers, but the green-backed heron uses bait, just like a human angler, to lure fish. It floats a piece of bait, which could be a feather, a fly, or a twig, and waits to strike at any fish that are attracted.

to spawn (lay eggs). One catfish called the dourada travels some 2,500 miles (4,000 km) within the river system itself, without ever swimming out to sea. Like brown bears waiting for salmon, human fishers wait to catch the passing fish at narrow places in rivers where there are rapids.

The migration of walking catfish in Southeast Asia is not limited to water—they can crawl across land and quickly colonize new habitats. This ability has made them a pest in the southern United States, where walking catfish have escaped from fish farms.

Tadpoles to Terrapins

Many freshwater habitats provide ideal breeding grounds for amphibians—animals such as frogs and salamanders that live partly

young are safer there. Sturgeons and other fish have similar migration patterns, but such fish cannot complete their journey when their route is blocked by a dam.

In the Amazon river system, several large catfish migrate from the tidal estuary region to the headwaters of the Amazon tributaries

The Blue Eye of Siberia

Lake Baikal in eastern Russia is not only the deepest lake in the world but also the oldest—perhaps 25 million years old or more. This has given plenty of time for evolution to produce unique species, and today many of the lake's inhabitants, from worms to fish and crustaceans, live nowhere else in the world. Ice-covered in winter, Baikal has deep currents that ensure there is plenty of oxygen even at the lake bed, 1 mile (1.6 km) below the surface. The lake also boasts the world's only freshwater seal: the Baikal seal, or nerpa (right).

In summer, millions of water-living insect larvae, such as those of mayflies and midges, emerge from rivers and ponds as winged adults. Flying predators such as bats, swifts, and swallows (below) swoop in, catching and eating them in midair.

in water and partly on land. The larvae of most amphibians have gills and live underwater, while the adults have lungs and legs and can walk on land—though they need to keep their skin moist. Amphibians breathe through their damp skin, and the skin of many species contains poison glands that ward off predators. Almost all amphibians are carnivores (they eat other animals).

Compared to other parts of the world, North and Central America have an unusually large number of salamander species. Salamanders are amphibians with long bodies and tails. Though many are land-living as adults, some species have returned to a completely aquatic existence. The mudpuppies and sirens, for example, keep their gills even as adults. The world's largest salamanders are the giant salamanders of rocky, fast-flowing Chinese and Japanese rivers. They can grow up to 5 feet (1.5 m) long. These night-active giants live permanently in water and can eat prey as big as rats and turtles.

Few frogs live entirely in water as adults, but their tadpoles are an important part of freshwater food chains—insects such as water beetles often eat them, for instance. Many tadpoles are filter feeders that eat plankton,

while others specialize in scraping algae from rocks, or extracting food from mud. A few aggressive species sometimes turn into meat eaters and eat other tadpoles.

In contrast to amphibians, reptiles (lizards, snakes, and relatives) lay eggs that are protected by hard shells and can hatch on land. Despite this, many reptiles have returned to a freshwater existence—though they still leave the water to lay eggs. Although some snakes and lizards hunt in water, the most typical freshwater reptiles are turtles and crocodiles. Freshwater turtles eat both plant and animal food and usually favor weed-filled ponds and slow-moving rivers. Among the most aquatic are the softshell turtles, which have flat, pancake-shaped bodies. These fast swimmers can breathe underwater through their skin and rarely come onto land. The world's largest freshwater turtle is the alligator snapping turtle of the southern United States, a massive creature that attracts prey by wiggling a wormlike projection on its tongue.

The bullfrog, like most frogs, spends most of its time motionless. Any passing small animal risks being caught by its tongue and pulled into its cavernous mouth.

A flash of blue is the last thing a fish sees when caught by a Eurasian kingfisher diving from a waterside perch. There are more than forty species of kingfishers.

The 23 species of crocodilians (crocodiles, alligators, caimans, and gharials) are top predators of rivers and lakes in warmer regions. A crocodile is a caring mother—she guards her buried eggs, digs them out when they hatch, and then gently carries the babies down to the water's edge in her mouth. One unusual species is the gharial of Indian rivers, a large but endangered crocodilian whose long, narrow snout is ideal for catching fish. It leaves the water only to bask and nest.

Birds and Beasts

Many birds take full advantage of the resources offered by rivers and lakes. Water's-edge birds include storks, herons, and other wading species, many of them equally at home in wetlands or by the seashore. On the open water, ducks and other floating birds use rivers and lakes not only for food, but as a

 # River Dolphins

Five species of river dolphins live in the great river systems of Asia and South America. They have poor eyesight and probably hunt mostly by bouncing sound waves off prey (a technique termed echolocation). They use their flexible neck to maneuver around underwater obstacles such as fallen trees. The Asian species that live in the Indus, Ganges, and Chang (Yangtze) rivers are among the most endangered mammals in the world.

haven from land-based predators. Some ducks eat plants, but others, such as mergansers, are specialized fish-hunters that use serrated (sawtooth-edged) bills to catch slippery prey.

Other birds swoop down on unsuspecting fish from the air. Large birds of prey such as ospreys and fish eagles sweep across the surface and grab fish with their talons. In Africa, a fish-eating owl does the same thing. In rushing streams around the world, a small songbird called the dipper has a different strategy: It simply walks along the bottom of streams in search of insect prey.

Among the mammals, various species have returned to a watery way of life. The most aquatic (though still air-breathing) freshwater mammals are those descended from sea mammals. They include river dolphins and the freshwater manatees of the Amazon. Other aquatic mammals, such as otters, frequently come onto land, despite having webbed feet and closable nostrils to help them underwater. Otters are playful mammals that are superb underwater hunters. Using their tails for propulsion, they are quick and nimble enough to chase and catch fish. The largest species—the giant otter of the Amazon—grows to 6 feet (1.8 m) in length. Heavily hunted by people, this otter is now the focus of several conservation efforts.

Smaller mammals, such as water shrews, also live in streams and ponds, and there are even several fishing bats that are able to grab their prey from above. At the other end of the scale, the hippopotamuses of Africa spend much of their lives in water, though they climb onto land to graze, usually at night. Despite their gentle appearance, hippopotamuses are dangerous animals, liable to charge and attack when they feel threatened. They kill hundreds of people in Africa every year.

The Platypus

The platypus of Australia is one of the most unusual mammals in the world. Growing up to 20 inches (50 cm) long, it hunts underwater prey in rivers and creeks, often at night. Uniquely for a mammal (but like many fish), the platypus, using its sensitive bill, can detect tiny electrical signals given off by its prey. Equally unusually, it lays eggs instead of bearing live young. It lays them in the safety of a long burrow that it digs into the riverbank.

Rift Valley Lakes

The lakes in the Great Rift Valley region of Africa are surrounded by biomes ranging from desert to rain forest. Hippos, crocodiles, vast flocks of flamingos, and an array of colorful fish are some of the lakes' varied residents.

 ## Fact File

▲ The Rift Valley lakes of East Africa formed 3–7 million years ago when upheavals in Earth's crust created a chain of deep valleys and mountains.

▲ The waters of lakes Tanganyika and Malawi are in separate layers that do not mix. Beneath the surface layer there is no oxygen and the water is saturated with toxic hydrogen sulfide, so no animals can live in their depths.

▲ At 4,700 feet (1,435 m), or nearly 1 mile deep, Lake Tanganyika is the second deepest lake in the world, after Lake Baikal in Russia.

 ## Victorian Fish

Lake Victoria, the largest lake in Africa, lies between the two arms of East Africa's Great Rift Valley. It is much shallower than lakes Tanganyika and Malawi, which lie in the Rift Valley itself. Although it was formed only 200,000 years ago, Lake Victoria, like lakes Tanganyika and Malawi, supports hundreds of species of fish. Most are members of the cichlid fish family and are found nowhere else. It is a mystery how so many species evolved in so short a time.

Fishing has always been important to the local human population, and to improve the catch of fish, the lake's residents introduced a large predatory fish in the 1950s. Called the Nile perch, this fish has since eaten its way through the native cichlids. The lake has also suffered from pollution by sewage and fertilizers from farms. This, together with modern fishing techniques and hungry Nile perches, has led to the tragic extinction of perhaps half of Lake Victoria's 500 or so native fish species.

Fishing communities line the shores of Lake Victoria. The people fish either from large trawlers or from canoes (below). They often equip the canoes with sails.

Above: Just one of the hundreds of amazing cichlid fish species living in Lake Malawi, this one broods its young in its mouth.

1. Nile River
The source of the Nile River is a stream flowing into Lake Victoria. When the river leaves the lake, it flows northward for 4,000 miles (6,400 km) to the Mediterranean Sea.

2. Great Rift Valleys
Forces deep inside Earth created this split down East Africa. Water collected in the deepest valleys and formed the Rift Valley lakes.

3. Lake Turkana
Until two million years ago, this lake was freshwater and connected to the Nile. Now its soda waters have a rich growth of algae that feeds shrimps that, in turn, feed flamingos.

4. Lake Victoria
Africa's largest lake is also the second biggest freshwater body in the world by area after Lake Superior, although its greatest depth is only 250 feet (80 m).

5. Lake Kivu
This lake was formed when lava dammed a river valley around 20,000 years ago. In January 2002, lava from the Nyiragongo volcano flowed into the lake, after first devastating the nearby city of Goma.

6. Eastern Rift Valley Lakes
Some of these lakes are freshwater, some are salty, and others are soda lakes. They are home to millions of flamingos, which move between the lakes to feed and breed. Lakes Manyara and Nakuru are both national parks.

7. Mara River
Every year millions of migrating wildebeest cross this river. Crocodiles hide in the water at crossings, waiting for a meal.

8. Rubondo Island National Park
This island in Lake Victoria is best known for its rich bird life. An introduced population of elephants also lives there.

9. Lake Tanganyika
This deep lake has a relatively small human population living on its shores. Its deepest point lies far below sea level and is also the lowest point on the African continent. It is home to hundreds of cichlid fish species.

10. Uwanda Game Reserve
This reserve includes most of Lake Rukwa, a salt lake noted for its bird life and a large population of crocodiles.

11. Lake Malawi
Lake Malawi contains more fish species than all of Europe's and North America's rivers and lakes put together. Malawi's 600 or more cichlid fish include some that eat only fish scales and others that live by nibbling the fins of other fish. Pollution and overfishing are beginning to put pressure on the lake's unique natural resources.

Rivers, Lakes, and People

Rivers and lakes have always been vitally important to people. Since earliest times, we've used them for drinking, washing, getting around, finding food, irrigating crops, and dumping waste.

Bonds between people and the freshwater biome are often intimate. In prehistoric Europe and elsewhere, people built artificial islands on lakes to live on for protection, while in eastern Asia today, the practice of living on houseboats or on stilt houses by the water's edge continues. Rivers also play an important role in spiritual and religious beliefs—the Ganges River in India, for example, is sacred in the Hindu religion.

Many of the world's rivers flow through landscapes transformed by centuries of human activity. In many cases, people have altered river and lake ecosystems drastically—sometimes on purpose, at other times accidentally. These changes have often turned out badly; freshwater biomes are probably the easiest for humans to damage.

People in Guangxi Province, southwest China, use diving cormorants to catch fish for them. Hoops around the cormorants' necks prevent them from swallowing large fish. The people extract the fish for their own use but allow the cormorants as much as they can eat at the end of the day's fishing.

In large lakes, freshwater fisheries reach an industrial scale. Trawlers on Lake Malawi (above) haul in catches of a variety of cichlid fish that are popular as food in Malawi, Mozambique, and Tanzania.

Fishing and Fisheries

For thousands of years, people have relied on freshwater fish as a source of food, especially for protein. Societies around the world have fished using hooks, spears, nets, and other methods. They also developed ingenious methods for trapping fish, ranging from baskets with funnel-shaped entrances left in the river, to specially dug sideponds that trapped fish when water levels fell. Sport fishing with a rod and line dates back at least to the ancient Romans, who even made jokes about it. In the United States, fly-fishing anglers have been one of the main pressure groups fighting to improve the quality of the nation's river and lake waters.

In modern times, catching freshwater fish has become a major commercial activity in places such as the Great Lakes, the Amazon, and the Mekong River in Southeast Asia. In these cases, a whole fishing industry, or fishery, has sprung up. Sometimes, new species of fish have been introduced in an attempt to make the fishing better—not always with the desired results. Like fishers at sea, freshwater fishers using modern methods tend to catch more fish than nature can put back. Small-mesh nets, for example, trap young fish before they've had a chance to breed. Sometimes, fishers even catch fish by the simple, though ecologically unfriendly, trick of throwing explosives into the water.

Reshaping Nature

For thousands of years, people have diverted and controlled river waters. There are many reasons for doing this: obtaining water for irrigation, public water supplies, or industry; creating inland routes for ships; controlling floods; or using water to provide energy. In the last hundred years, such interventions have grown in scale. Some projects, such as the flood-control measures on the Mississippi, have become controversial in recent years, with many people arguing that they make the flooding worse, not better.

Most large schemes involve building dams on rivers. For example, the upper Mississippi was dammed to aid river navigation in the late 1800s, with the program greatly extended in the 1930s. Large dams and reservoirs now interrupt many of the world's major rivers systems, such as the Volga in Russia and the Indus in Pakistan.

Dams can harm the natural environment in several ways. First, they interrupt the natural flooding of a river and the patchwork of wetlands that the floods create. Turtles, for example, may lose the sandbanks that they need to lay eggs on. Dams also prevent migrating fish from swimming upstream to their breeding grounds—damming the Volga greatly disrupted the life cycle of the sturgeon that produce Russia's famous caviar, for example. And the lake behind the dam may be too warm or too low in oxygen for some fish to survive at all.

In the 1980s, the Soviet Union devised the world's most grandiose river-control program. This would have involved preventing the country's large, northward-flowing rivers from reaching the Arctic Ocean, and diverting them so that they irrigated southern,

Aquaculture

Aquaculture is underwater agriculture—growing aquatic animals or plants under controlled conditions, usually for food. Shrimps and crayfish are sometimes farmed, but the largest branch of freshwater aquaculture is the farming of fish. It has long been practiced in China, which produces two-thirds of the world's farmed fish. Production is often boosted by keeping several species together—one fish feeds on underwater plants, another on small mud-dwelling creatures, and so on. Alternatively, fish can be farmed in rice paddies, as they are in Thailand (right).

Fish farming is also a major activity in Mississippi and other southern states, where catfish are the main product. Aquaculture is a more efficient way of producing high-protein food than, say, cattle ranching, and will probably become increasingly important in the future.

The Aral Sea Disaster

An example of how people can alter ecosystems is the notorious case of the Aral Sea disaster in Central Asia. The Aral Sea was once the world's fourth largest lake and supported a major fishing industry. During the 1960s, the authorities in the Soviet Union diverted much of the river water feeding this salty lake into irrigation projects. Factories and fields of cotton developed in the valleys of the Amu Darya and Syr Darya rivers.

Downstream, though, the Aral Sea (below right) shrank in area by more than half, the concentration of salt tripled, the fish population collapsed, and fishing ports now lie miles from the water's edge, their rusting boats (above right) stranded on dry land. Even the irrigation projects have started to fail, because of salt and pesticide contamination and resulting human health problems.

industry | fishing port (derelict) | cotton fields

City authorities dye the Chicago River green each year to mark St. Patrick's Day. The dye is not toxic; in fact it is usually used to trace illegal pollution sources. In large quantities, though, the dye itself can be a pollutant.

grain-producing lands instead. Apart from disrupting the natural ecosystem, this huge project might have unbalanced the formation of ice in the Arctic Ocean, with potentially grave effects on the world's climate. The idea was eventually dropped, but for political and economic reasons, not environmental ones.

Pollution and Water Quality

Because rivers are the land's natural drains, much pollution eventually ends up in freshwater, wherever it starts off. Sources of pollution might be specific places, such as factories, mines, sewage outfalls, accidents, and spillages. A single road tanker spilling its load can undo years of work spent restoring a river to a healthy state. The location of specific sources can often be identified easily. But pollution also comes from nonspecific places including farmland and city streets. While the effects of this pollution are not sudden or obvious, they are no less serious.

Another kind of pollution is thermal—in other words, warm water. Usually discharged from power plants, which need water for cooling purposes, warm water contains less oxygen and can make it impossible for cold-loving species such as trout to live nearby.

Some pollution is difficult to detect by chemical tests. Instead, scientists can assess the level of pollution indirectly by examining the animal life in a river or lake. Some small animals, such as bloodworms, can tolerate pollution, while others, such as stonefly larvae, are more sensitive and survive only in the cleanest waters.

The oldest type of water pollution is sewage, which has been a problem ever since people started living in large cities.

Untreated sewage dumped into water makes it smell and look horrible; it feeds bacteria that use up oxygen, causing fish to suffocate; and it spreads diseases, including cholera and typhoid. In rich countries, wastewater treatment plants turn sewage into clean water, but in many poor parts of the world, untreated sewage is still a major problem.

Another well-known pollution problem is acid rain, which is caused by gases such as sulfur dioxide, released when fossil fuels are burned. These gases can blow thousands of miles on the wind and dissolve in rainwater, making it acidic. Sometimes soil or water in the ground can neutralize this acid, but not always. In Europe, thousands of lakes in Norway and Sweden have lost fish due to acid rain, most of it originating in other countries. It is not always the acid itself that causes damage—acid rain also dissolves aluminum from soils, and the aluminum is poisonous to many creatures.

Artificial Lakes

In some parts of the world, the largest lakes are not natural ones but artificial reservoirs created by damming rivers. Several of the biggest are in Africa—including Lake Nasser on the Nile, Lake Kariba (above) on the Zambezi, and Lake Volta in western Africa. Studying how these lakes develop gives us clues to how natural lakes develop. Some of the dammed lakes initially went through a low-oxygen phase, as drowned trees decayed, but in time most have established successful fish populations. Lake Nasser now provides local people with a rich source of fish, but fishers at the other end of the Nile River were not so lucky. The lake's dam stopped nutrients from reaching the Nile Delta, killing the sardine population that thrived offshore.

Freshwater habitats are often polluted by fertilizer chemicals from farms. Crops are frequently treated with artificial fertilizers (especially nitrates and phosphates) to help them grow. These fertilizers dissolve in rain and can end up draining into nearby rivers, lakes, or ponds, where they stimulate the growth of aquatic algae. The water turns a murky green and its oxygen level falls, causing fish and other animals to suffocate. This process is called eutrophication, and it sometimes occurs naturally. However, modern agriculture has made it much more common.

Human wastes and phosphates from detergents also cause eutrophication, and the latter are now controlled in the United States. The problem became famous when Lake Erie was reported to be dying in the 1960s due to eutrophication. Since then, the situation has improved. One natural method used to control the problem is to let the water pass through a marshy wetland first.

Although polluted, the Ganges River in India is a holy river for Hindus. Hundreds of thousands of pilgrims come to take part in annual bathing festivals.

The growing wetland plants take up some of the nutrients, and bacteria turn some of the nitrates into harmless nitrogen gas.

Hidden Dangers
It used to be said that "dilution is the solution to pollution." In other words, dilute the pollution with enough water, and it won't be a problem anymore. Unfortunately, this is not true for certain pollutants. Some chemicals not only hang around in the environment without decaying, but also accumulate in animals' bodies until they reach a concentration millions of times that in the surrounding water. This process is called bioaccumulation.

Bioaccumulating pollutants include the so-called heavy metals—poisonous metallic elements such as lead, cadmium, and mercury, which are often released by industry or mines. In Europe, the Rhine River delta is full of heavy metals that washed down from industries upstream over several decades. Mercury can also be released into the air from smokestacks and be blown to rivers and lakes in distant places. It is currently a major

Major rivers such as the Mississippi reach deep into the world's continental landmasses and become convenient highways for people to transport bulky goods.

pollution problem in North America—advisory notices have been issued for hundreds of lakes, telling people not to eat the fish they catch there.

Some pollutants bioaccumulate by dissolving in the fatty tissues of animals. These include a family of chemicals called PCBs and the notorious insecticide (insect-killing chemical) DDT, widely used in U.S. agriculture in the 1940s and 1950s, and now used to control malaria-carrying mosquitoes. Animals at the top of the food chain are most at risk of poisoning—including ourselves, if we eat contaminated fish.

PCBs have some strange effects, even at low concentrations. Once inside an animal's body, they act like female sex hormones, turning a male into a female, or into an animal that is half male and half female. This effect is common in fish and also happened to a population of alligators in polluted Lake Apopka, Florida. PCB pollution may even be affecting people—some scientists think it is why men today produce only half as many sperm as they did fifty years ago.

New Arrivals

People have altered freshwater ecosystems by introducing animals and plants from other parts of the world. Nonnative aquatic plants are some of the worst weeds in the world, and introduced animals often have a major impact on existing wildlife.

It has been estimated that, over the past 50 years, at least 150 species of freshwater fish have been deliberately moved around the world to improve fishing in new areas. Sometimes little disruption occurs, but in other cases there can be damaging effects. The most notorious case is the Nile perch, a huge fish-eating African species that people introduced to Lake Victoria in Africa during the 1950s. Many of the unique fish species that once lived in the lake are now extinct because of this alien predator. Today, the local fishers catch a lot of Nile perch, but

Murky Waters

Rivers vary in how much sediment (mud, sand, gravel, or boulders) they carry, and the animals that live in them reflect this. So when people change the sediment entering a river, they change its wildlife. When settlers first plowed the land around the Mississippi for crops, soil was washed into streams, smothering native species, such as freshwater mussels, and clogging up gravel beds where fish lay their eggs. When it reached the coast, the Mississippi's sediment once dispersed into the sea and over thousands of years it built the Mississippi Delta (right). Today, the river is artificially channeled through its delta. The sediment is no longer dispersed, so the delta is now shrinking.

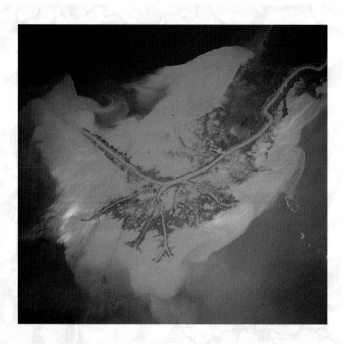

The Huang He (Yellow River) is a turbulent and unpredictable river that freezes over in its middle section each winter. Nevertheless, people have used it for travel for thousands of years, and it flows near some of China's oldest cities

it is an oily fish and must be smoked to preserve it. So people have to cut down many trees for firewood, resulting in more damage to the environment.

Fish and other animals are sometimes introduced to new habitats accidentally. The sea lamprey—a sucker-mouthed fish that rasps the flesh of trout and other fish— probably got into the Great Lakes via canals, which provided a bypass around Niagara Falls. Once there, the lamprey virtually wiped out the lakes' trout; today, it is finally being brought under control. The zebra mussel is another notorious invader. It and many other recent arrivals probably got to the Great Lakes via the ballast water that ships take in to weigh themselves down. The salty water of the Caspian Sea has been invaded by a jellyfish-like animal called a comb jelly, which somehow got there from North American coastal waters.

Above: Fishers on the Congo River display a Nile perch, which lives in several major African rivers. When introduced to lakes, it usually wreaks ecological havoc.

 # Striped and Dangerous

The zebra mussel is one of the most damaging animals to invade North American freshwaters. A native of the Caspian Sea area in Central Asia, it reached Great Britain by the 1830s and Lake Erie in 1988, almost certainly because its tiny swimming larvae were carried there by ocean-going ships. It has since spread throughout the Great Lakes and into the Mississippi river system. It is the only freshwater mussel that can attach to hard surfaces, holding on by tough, nylonlike threads. Unfortunately, hard surfaces include water pipes, which it clogs, and the shells of native mussels, which it can smother. It is so efficient at filtering food from water that it has made the water in parts of Lake Erie clearer than before—but there is now less food for fish. Recently, though, there have been signs that the zebra mussel is declining in number, perhaps because its predators, such as diving ducks, are learning to attack it.

Huang He and Chang

China has a very long history, and its two greatest rivers, the Huang He (Yellow River) and the Chang (Yangtze), have been bound up with the fortunes of its peoples for thousands of years.

The Three Gorges Dam

The massive Three Gorges Dam now being built on the Chang is an extremely controversial project. The developers have several aims: to use turbines in the dam to generate electricity, to help ships pass up the river, to control floods, and to allow diversion of river water to other areas. The spectacular Three Gorges (below) will not be submerged, because their walls are up to 4,000 feet (1,200 m) high and the river level will rise much less than this. The area to be flooded, however, has been home to people for thousands of years, and flooded graveyards and mine refuse may cause pollution—in addition to the cultural and human loss due to ancient towns and cities being flooded. The dam will also trap pollution from the city of Chongqing, lying upstream.

1. Tibetan Plateau
The highest plateau in the world is the source of many of Asia's greatest rivers, including the Ganges and the Indus besides the rivers of China.

2. Tanggula Shan
The Chang (Yangtze) River begins as meltwater from snow and glaciers on the high Tanggula Shan mountains. The water forms streams that flow across the Tibetan Plateau.

3. Brahmaputra River
This great river starts in Tibet. The Chinese government has considered diverting some of its waters to the Chang.

4. Mekong River
A river very rich in animal life, especially its fish species. Chinese plans to divert some of its waters to the Chang could greatly impact countries farther downstream.

5. Yunnan-Guizhou Plateau
Rivers of the Chang system cut deep gorges through this limestone region and also flow underground through caverns. In these isolated rivers live mustached toads, firebelly newts, seven species of cave fish, and the baggy-skinned Chinese giant salamander.

6. Loess Deposits
The Huang He passes through a landscape built of loess (dusty wind-deposited sediment). The river picks up huge amounts of loess, giving the river a yellow color—hence its name.

7. Lakes Region
The lakes of this populous, fertile region help to absorb the Chang's floodwater. Wildlife swims into the lakes during floods and back to the river in the dry season.

8. Huang He (Yellow River)
The Huang He sometimes causes enormous floods. The Chinese have been trying flood controls for thousands of years.

9. Grand Canal
More than 2,000 years old, the Grand Canal joins the Chang with the Huang He. There are plans to use it to supply more water to north China.

10. Lower Chang River
Rare and endangered animals live in the lower Chang, including the Chinese river dolphin, the finless porpoise, and the Chinese alligator.

11. Old Huang He Delta
In A.D. 1194 during a great flood, the Huang He altered its course and flowed into the sea farther south for the next 700 years. In the 19th century, it altered its course back again.

12. Shanghai
China's largest city lies near the mouth of the Chang. If water is diverted northward from the Chang, seawater may seep into the Shanghai area to replace the freshwater, affecting water supplies.

Above: Both Chang and Huang He carry a heavy load of sediment. Here, the Chang dumps yellow loess at its mouth in the East China Sea.

Below: The Huang He is a muddy torrent in the wet season, but in some years it dries up before reaching the sea.

Fact File

▲ At 3,915 miles (6,300 km) long, the Chang is Asia's longest river and the third longest in the world, just beating the Mississippi river system.

▲ Large cities have never been built on the banks of the Huang He for fear of its notorious floods.

▲ The sediments laid down by the Huang He often cause its river channel to be higher than the surrounding countryside—one reason why its floods can be so disastrous.

The Future

While people pour pollutants into lakes and rivers and take the water to supply homes and industry, can the wildlife in the water survive?

Many people would be saddened if river dolphins, giant otters, and other spectacular freshwater creatures were to become extinct in the near future as a result of human carelessness. On top of that, there's our own future to think about. Worldwide, the human population is in great danger of running out of freshwater during the coming century. For that reason alone, we need to look after our rivers and lakes as a vital resource. Although all the world's biomes are under threat, rivers and lakes are very much at risk when it comes to the inevitable clash between the need for economic growth in the traditional sense and the long-term future of ourselves and the planet.

Be Very Afraid...

There are lots of reasons to feel gloomy about the future of rivers and lakes. In the developed world, much ecological damage has already been done—rivers including the Mississippi and the Colorado no longer flow freely, while ecosystems such as wetlands that they once supported have now disappeared. One of Africa's largest lakes, Lake Chad, has recently shrunk to a fraction of its former size due to overuse of its waters. In other parts of

Like all hydropower plants, this one on the Niagara River gives cheap, clean, renewable power. But hydropower projects can devastate freshwater ecosystems.

Northern part

causeway

Southern part

Salt Lake City

N

 ## Salt Lake Blues

Utah's famous Great Salt Lake (left) has been in trouble in recent years. It is so salty that fish cannot normally live in it, but the brine shrimp (inset) thrives there and supports huge populations of migrating birds, as well as a commercial industry that scoops brine-shrimp eggs out of the water. In 1959, however, a causeway was built across the middle of the lake to carry a railroad, with only a few small gaps left for water to get through. This interfered with the mixing of water in the lake. The north part has since become even saltier—too salty for the brine shrimp—and algae now often turn it the color of strawberry milk. The southern part, by contrast, receives most of the river input and so is now more dilute, and the brine shrimp is suffering there, too. Environmentalists fear that the bird populations using the lake may decline unless better mixing of the two halves of the lake is restored.

the world, huge dam projects are still being started for hydropower or irrigation, such as the Three Gorges Dam being built on the Chang River in China. Such dams can lead to disruption of ecosystems and displacement of thousands of local people. Badly designed or implemented irrigation projects can also result in pollution and salt-poisoning of soil.

The Colorado and the Rio Grande are examples of international rivers whose resources are claimed by more than one country—in these cases the United States and Mexico. Many large rivers flow through a number of different countries, and so there is huge potential for disputes—even wars—if an upstream country starts taking too much water. In Africa, for example, the Nile, Niger, and Zambezi rivers are all international rivers that are potential flash points for the future.

New chemicals are also being brought into use all the time—and it may be years before we detect the damage some of them do to the environment. Besides, existing pollution already threatens many freshwater fisheries. Global warming, too, is likely to alter the pattern of the world's rainfall, resulting in unpredictable changes to river flow, river flooding, lake levels, and lake saltiness.

Increasing international trade means that, inevitably, more alien species will find their way from one continent to another, disrupting natural ecosystems. There's also a continuing trade—both legal and illegal—in freshwater animals themselves. For example, in Southeast Asia, rare turtle species are being collected in large numbers and exported to China for food, or for use in traditional Chinese medicine.

Underlying many of these problems is the fact that the world human population is still growing very fast. Few national governments have taken action to change population trends. The government of China, however, operates a strict policy that

Ordinary garden ponds can do much to provide a refuge for amphibian species threatened by the destruction of their natural habitats.

limits its citizens to having small families. Even with the existing human population, though, many people in poor countries have no access to safe drinking water and are in urgent need of better supplies. Furthermore, more of the world's people are living in cities. Due to the high-energy urban lifestyle, the average city dweller uses much more water in the home than a person living in the country.

Signs of Hope

There is a brighter side, however. To begin with, we're beginning to understand how freshwater ecosystems work, and so are more aware of the damage our actions might do. Today, for example, any major new dam proposal usually results in an international outcry at least. In the United States, the passing of the Clean Water Act in 1972 was a first step toward tackling the problem of pollution in the nation's rivers and lakes.

Strict legal protection has rescued the American alligator—hunted almost to extinction 50 years ago. But changes to its habitat may pose a new threat.

Local action has resulted in the cleaning up of many rivers—the Charles River in Massachusetts is a well-known example. Many older dams have also been dismantled in states such as Maine and California, allowing rivers to run freely again and enabling migratory fish such as salmon to swim up them to breed. Animals such as the American alligator have also benefited from new laws protecting them from exploitation.

In England, a homely but welcome development has been the growing popularity of garden ponds—these have done much to preserve populations of frogs and newts threatened by the destruction of pond habitats in the countryside.

There may be bright aspects to the long-term future, too. For example, much electricity currently comes from hydropower, which is cheap and clean but can disrupt freshwater ecosystems drastically. If we're lucky, even cheaper, cleaner power may one day become available from a process called nuclear fusion, which would produce far less pollution than today's nuclear power stations.

Lake Superior is at the heart of one of the world's largest freshwater ecosystems. As more people come to enjoy its wild beauty and its unrivaled leisure facilities, shoreline developments are damaging fish spawning grounds, amphibian habitats, and bird nesting sites.

Water Power to the People

In the United States, the federal Clean Water Act allows citizens to take action to clean up their own watersheds. The powers it gives center around the concept of a "total maximum daily load" of pollution. Called TMDL for short, this is the amount of pollution that any river or lake can take while still remaining healthy, wherever the pollution happens to come from. Using Clean Water Act legislation, citizens can monitor TMDL reports for their local watershed and force states and local governments to come up with an action plan to reduce pollution.

Glossary

algae (singular: alga): Simple, plantlike organisms that make food from sunlight like plants do but lack proper roots and leaves.

amphibian: An animal, such as a frog or salamander, that lives partly in water and partly on land.

aquatic: Aquatic animals and plants live in water for at least part of their life cycle.

asexual reproduction: A method, used by some animals and plants, in which mixing of genes between parents is not needed.

bacteria (singular: bacterium): Single-celled microorganisms; among the tiniest and simplest forms of life.

basin: A bowl-shaped lowland drained by a single river system.

biome: A major division of the living world, distinguished by its climate and wildlife.

camouflage: A natural disguise that makes animals or plants look like their surroundings.

carbon dioxide: One of the gases in air. Animals and plants produce carbon dioxide constantly.

cichlid fish: Any of the cichlid fish family. Many different cichlids live in the lakes of East Africa.

climate: The pattern of weather that happens in an average year.

community: A collection of organisms living in the same place, such as a stream or a lakeshore.

crustacean: An invertebrate, such as a crab, crayfish, or brine shrimp, often with ten pairs of jointed legs.

delta: A wide, typically Δ-shaped plain at the mouth of a river where sediment collects.

desert: A place that gets less than 10 inches (250 mm) of rain a year.

ecological: To do with the way organisms interact with one another and the environment.

ecosystem: A collection of living animals and plants that function together with their environment.

emergents: Plants, such as reeds and cattails, that grow through shallow water and into the air.

equator: An imaginary line around Earth, midway between the North and South poles.

estuary: A place where the tide mixes with freshwater from a river, usually in an inlet of the sea at the mouth of a river.

eutrophication: The process that happens when a river or lake becomes full of nutrients. It often occurs due to pollution by fertilizer.

evaporate: To turn into gas. When water evaporates, it becomes an invisible part of the air.

evolve: To change gradually over many generations.

fertile: Capable of sustaining plant growth. Farmers often try to make soil more fertile when growing crops.

fertilizer: An added substance, often an artificial chemical, that makes farmland fertile.

food chain: Scientists can place animals and plants into a series that links each animal with the plant or animal that it eats. Plants are usually at the bottom of a food chain with carnivores at the top.

freshwater: Water with only a very small amount of dissolved salt. Nearly all rivers are freshwater.

glacier: A river of ice.

global warming: The gradual warming of Earth's climate, thought to be caused in part by pollution of the atmosphere.

hormone: A messenger chemical in a plant's or an animal's body.

hydropower: Electricity generated by water turning turbines.

ice age: A period in history when Earth's climate was cooler and the polar ice caps expanded. The last ice age ended 10,000 years ago.

invertebrate: An animal with no backbone, such as a mussel.

irrigation: The use of artificially channeled water to grow crops.

mammal: An animal, usually furry, that feeds its young on milk.

migration: A journey made by an animal to find a new home.

nutrient: Any chemical that nourishes plants or animals, helping them grow.

oxygen: A gas in the air, needed by animals and plants to release energy from food.

parasite: An organism that lives inside or on another organism and harms it.

phytoplankton: Plantlike members of the plankton.

plankton: Organisms that drift along with ocean or lake water, mainly near the surface.

plateau: An area of relatively flat land higher than its surroundings.

pollen: Dustlike particles produced by the male parts of a flower.

pollination: The transfer of pollen from the male part of a flower to the female part of the same flower or another flower, causing the flower to produce seeds.

pollution: Contamination of the environment by artificial substances such as sewage and chemicals from farms and industry.

predator: An animal that catches and eats other animals.

protein: One of the major food groups. It is used for building and repairing plant and animal bodies.

rain forest: A lush forest that receives frequent heavy rainfall.

reptile: An animal such as a crocodile or turtle that usually has scaly skin and lays its eggs on land.

rhizomes: Horizontal stems used by a plant for spreading and for food storage.

river system: A river and all of its tributaries.

sediment: Any solid material (mud, silt, sand, gravel, pebbles, or boulders) carried by a river.

sexual reproduction: Any reproduction involving mixing of genes between parents.

species: A particular type of organism. Cheetahs are a species but birds are not, because there are lots of different types of birds.

tributary: A river or stream that joins a larger river or stream.

tropical: Within 1,600 miles (2,600 km) of the equator.

tundra: A cold biome of the far north, made up of treeless plains.

vapor: A gas formed when a liquid evaporates.

vertebrate: An animal with a backbone, such as a fish or human.

watershed: An area of Earth from which water drains toward one river system. A watershed includes a main river valley and all the tributary valleys. *Watershed* can also mean the area boundary.

wetland: A biome with water-logged soil. Swamps and marshes are part of the wetland biome.

zooplankton: Animal-like members of the plankton.

Further Research

Books

Burgis, Mary J. and Morris, Pat. *The Natural History of Lakes.* New York: Cambridge University Press, 1987.
Farndon, John. *The Wildlife Atlas.* New York: Reader's Digest, 2002.
Morris, Neil. *Rivers and Lakes (Wonders of Our World, No. 1).* New York: Crabtree, 1998.
Pringle, Laurence. *Rivers and Lakes (Planet Earth Series, 17).* New York: Time Life, 1985.

Websites

Wild World Global 200: www.nationalgeographic.com/wildworld/global.html
(Profiles of the 200 richest and most endangered wild regions, according to the World Wildlife Fund.)
Great Lakes Information Network: www.great-lakes.net/
Amazing Amazon Facts Page: www.amazon-ecotours.com/Amazon.htm
All Along a River—a site on rivers written by students: library.thinkquest.org/28022/?tqskip1=1&tqtime=0117

Index

Picture Credits